German Federal Agency for Nature Conservation (Ed.)

Biodiversity and Tourism

This report was compiled on commission of the German Federal Agency for Nature Conservation (BfN) and funded by the German Federal Ministry of the Environment, Nature Conservation and Nuclear Safety.
The report contains the opinion and views of those commissioned to compile it and must not be in agreement with the opinion of the parties commissioning/financing it.

Springer

Berlin
Heidelberg
New York
Barcelona
Budapest
Hong Kong
London
Milan
Paris
Santa Clara
Singapore
Tokyo

German Federal Agency for Nature Conservation (Ed.)

Biodiversity and Tourism

Conflicts on the World's Seacoasts and Strategies for Their Solution

With 45 Figures, and 26 Tables

Springer

German Federal Agency for Nature Conservation
Konstantinstraße 110
53179 Bonn

Report compiled by:
Office of Tourism and Outdoor-Recreation Planning
Friedbergstraße 19, D-14057 Berlin
Telephone (+49) 30/324 76 79, telefax (+49) 30/324 26 78
Project directors: Dipl.-Ing. Hartmut Rein and Prof. Dr. Helmut Scharpf

The legal portions of this report were compiled by:
Dr. Lothar Gündling, attorney-at-law
Sofienstraße 23, D-69115 Heidelberg
Telephone (+49) 6221/16 28 74, telefax (+49) 6221/16 28 50

Project administrator:
Georg Fritz, German Federal Agency for Nature Conservation
Konstantinstraße 110, D-53179 Bonn

English translation:
Philip Mattson, Heidelberg

ISBN-13: 978-3-642-64508-2 e-ISBN-13: 978-3-642-60689-2
DOI: 10.1007/978-3-642-60689-2

Die Deutsche Bibliothek - CIP-Einheitsaufnahme

Biodiversity and Tourism: Conflicts on the World's Seacoasts and Strategies for Their Solution /
German Federal Agency for Nature Conservation (Ed.). - Berlin; Heidelberg; New York; Barcelona; Budapest; Hong Kong; London; Milano; Paris; Santa Clara; Singapore; Tokyo: Springer 1997
German edition: Biodiversität und Tourismus

Softcover reprint of the hardcover 1st edition 1997

Typesetting and Layout: Graphischer Arbeitsraum Felix Dahms, Berlin

Foreword

It is in the best interest of all concerned that tourism become sustainable and environmentally compatible. This need for "sustainable development" is being recognised more and more by the responsible parties. Moreover, in the search for solution strategies the realisation is gaining ground that tourism must be viewed as a worldwide phenomenon whose development must be co-ordinated in a co-operative effort spanning regions and continents.

That the preservation of biological diversity also requires global co-operation has been confirmed by over 170 countries which have already acceded to the "Convention on Biological Diversity".

It is thus an important task to provide the foundations for joint action. Germany, one of the largest source countries of international tourism, must feel particularly obligated in this regard.

The report published here is the result of a research project commissioned by the German Federal Agency for Nature Conservation. The study pursued and examined the thesis that the Convention on Biological Diversity be used as a central instrument for arriving at international principles and regulations for combining nature conservation and tourism which could lead to a sustainable development of tourism. To further the discussion, the authors brought the study to a logical conclusion by working out a proposal for a "tourism protocol" additional to the existing Convention on Biodiversity. Such a protocol additional to the Convention would entail the stipulation of internationally binding implementation and regulations for achieving sustainable tourism. However, it would be premature to speak of a marked propensity as of now on the part of the contracting states of the Biodiversity Convention to take up such a proposal without delay. Nonetheless, the proposal must land on the table of the parties it is addressed to as soon as possible.

The analysis of conflict situations and the most severely impacted areas of the world clearly shows that there is considerable need for action to make the development of tourism compatible with the conservation of biological diversity.

The discussion of existent planning and legal solution strategies is not only evidence that even now elements of guidelines and regulations can be implemented, but is also intended to provide impetus for bringing these solutions to realisation.

May this report, with its information and stimulation of discussion, contribute to this common concern to find not only an economic future for tourism, but also an ecologically sound one.

Prof. Martin Uppenbrink
President
Federal Agency for Nature Conservation

Introduction

More than 170 contracting states have signed the "Convention on Biological Diversity". This biodiversity agreement is designed to establish a new global partnership for the protection of living nature. Despite increasing efforts undertaken by individual nation-states and the international community of states in the last two decades, the loss of biological diversity has increased at an ever faster rate. The destruction of habitats, their overexploitation and the discharge of harmful substances have triggered this decimation of resources. The reasons for this are in turn primarily the production and consumption patterns which negate the principle of sustainable management and use of biological resources. Tourism is directly involved in this problem, which is of such vital importance for the future of mankind in two ways.

First, a great number of studies conducted on a global scale have shown that in its various forms, tourism to a considerable degree shares responsibility for the loss of species and habitats. The false location of touristic amenities has contributed to this, as have the carrying out of leisure-time activities in sensitive areas or the introduction of harmful substances to ecosystems by tourists. On the other hand, it is precisely tourism which in many areas is highly significant in contributing to the securing of biodiversity. Thus, what has come to be called ecotourism has become one of the most important instruments of nature conservation. The biological resources are being protected because their touristic value has been discovered.

Before the backdrop of a continued global expansion of tourism development which is furthermore concentrated more and more on areas whose species and habitat diversity are particularly significant, the complex of problems involving tourism and biodiversity should be given special attention within the framework of implementing the Convention.

The German federal government therefore decided to have the information on the general subject of tourism/biodiversity available to date be compiled and analysed in systematic form. This study is not only to document the existing pressure stemming from the problem by citing empirical material, but also systematically present the measures taken to date worldwide to solve the problem and as far as possible comment on them critically. The study is concentrated on coastal regions, as such regions are worldwide places where tourism is particularly concentrated and at the same time constitute areas of particular significance for the diversity of ecosystems and species.

These studies concerned with biodiversity and tourism form the background for the draft of a tourism protocol or corresponding guidelines which could contribute in this area to lending the Convention on Biological Diversity more concrete form.

It need not be described here in detail what difficulties were involved in fulfilling the task as commissioned in a very short period of time – with funding which ruled out any field research from the beginning. The result can thus only be a first, albeit, in the authors' view, comprehensive overview of the problem which affords valuable strategies for further regional steps.

In the legal section possibilities will be suggested for further developing international law in the direction of implementing sustainable tourism projects.

Table of Contents

Appendix

List of Contributors

R & D Project
"Sustainable Tourism Programmes for Coastal Regions"
(UFOPLAN Project FKZ 101 02 12) – Final Report –

compiled on commission of the German Federal Ministry for the Environment, Nature Conservation and Nuclear Safety, represented by the German Federal Agency for Nature Conservation, FG II 5.4, Konstantinstrasse 110, D-53179 Bonn

Büro für Tourismus und Erholungsplanung (BTE)
Office of Tourism and Outdoor-Recreation Planning
Friedbergstr. 19, D-14057 Berlin
Telephone +49 30/324 76 79
Telefax +49 30/324 26 78

Project directors:
Dipl. Ing. Hartmut Rein
Prof. Dr. Helmut Scharpf

Final compilation:
Dipl. Ing. Anja Maschewski
Dipl. Ing. Wolfgang Strasdas

Contributions by:
Cand. Ing. Matthias Schmidt
Cand. Ing. Eva Ebsen
Cand. Ing. Heike Hessler
Cand. Ing. Frank Kosching
Cand. Ing. Petra Ruth
Cand. Ing. Jan Schubert
with a contribution on Spain by
Dipl. Ing. Thomas Thierschmann

The legal portions of this report were compiled by:
Dr. Lothar Gündling, attorney-at-law
Sofienstr. 23, D-69115 Heidelberg
Telefon +49 62 21/16 28 47
Telefax +49 62 21/16 28 50

Graphics:
© 1996 by Felix Dahms, Dipl. Des., image/art author no. 64 08 61
Merseburger Str. 1, D-10823 Berlin
Telephone +49 30/78 70 47-00

English translation:
Philip Mattson

Berlin/Heidelberg, March, 1996

List of Abbreviations

ADAC	Allgemeiner Deutscher Automobil-Club (German General Automobile Club)
AGÖT	Arbeitsgemeinschaft Ökotourismus (Ecotourism Study Group)
AMAP	Arctic Monitoring and Assessment Programme
BfN	Bundesamt für Naturschutz (German Federal Agency for Nature Conservation)
BMFT	Bundesministerium für Forschung und Technologie (German Federal Ministry for Research and Technology)
BMP	Baltic Monitoring Programme
BMU	Bundesministerium für Umwelt, Naturschutz und Reaktorsicherheit (German Federal Ministry for the Environment, Nature Conservation and Nuclear Safety)
BMZ	Bundesministerium für wirtschaftliche Zusammenarbeit und Entwicklung (German Federal Ministry for Economic Co-operation and Development)
BNatSchG	(German) Federal Nature-Conservation Law
BTE	Büro für Tourismus- und Erholungsplanung (Office of Tourism and Outdoor-Recreation Planning)
CAMP	Coastal Area Management Programme
CCE	Comisión de la Comunidad Europea
CDPE	Committee for the Protection and Management of the Environment and Natural Habitats
CDT	Commonwealth Department of Tourism
CEC	Commission of the European Community
CORAL	Coral Reef Alliance
CPD	Centers of Plant Diversity
CTO	Caribbean Tourism Organization
DC	Development co-operation
DEHOGA	Deutscher Hotel- und Gaststättenverband (German Hotel and Restaurant Association)
DFV	Deutscher Fremdenverkehrsverband (German Tourism Association)
DGF	German Leisure-Time Association
DP	Development plan
DRV	Deutscher Reisebüroverband (German Travel Agents' Association)
ECNC	European Centre for Nature Conservation
ECTWT	Ecumenical Coalition on Third World Tourism
EEA	European Environment Agency
EIA	Environmental-impact assessment
EIS	Environmental-impact study
ESCAP	Economic and Social Commission for Asia and the Pacific
EU	European Union

EUCC	European Union for Coastal Conservation
FAO	Food and Agriculture Organization
FFHD	Fauna-Flora-Habitat Directive
FNNPE	Federation of Nature and National Parks of Europe
FÖNAD	Föderation der Natur- und Nationalparke Europas (Federation of European Nature and National Parks)
GEF	Global Environmental Facility
GIS	Geographic-information system
HELCOM	Helsinki Commission
ICEC	International Council for the Exploitation of the Sea
ICM	Integrated Coastal Management
ICONA	Institute for the Conservation of Nature
ICT	Instituto Costarricense de Tourismo
ICZM	Integrated Coastal Zone Management
IEEP	Institute for European Environmental Policy
ISEP	International Society for Environmental Protection
ISSG	Irish Sea Study Group
IUCN	International Union for Conservation of Nature
MAB	Man and Biosphere Programme
MAP	Mediterranean Action Plan
MEDPOL	Mediterranean Pollution Monitoring and Research Programme
NGO	Non-government organisation
NP	Nature parks /Parces naturales
OECD	Organization for Economic Cooperation and Development
Ö.T.E.	Ökologischer Tourimus in Europa (Ecological Tourism in Europe)
PAP	Priority Action Programme
PATA	Pacific Asia Travel Association
PORN	Plan de Ordenación de Recursos Naturales
PRUG	Plan Rector de Uso y Gestión
RAM	Rapid-assessment matrix
RPP	Regional-planning procedure
TES	The Ecotourism Society
UNCED	United Nations Conference for Environment and Development
UNESCO	United Nations Educational, Scientific and Cultural Organization
UNDP	United Nations Development Programme
UNEP	United Nations Environmental Programme
UNEP IE	United Nations Environmental Programme Industry and Environment
USAID	United States Agency for International Development
WTTC	World Travel and Tourism Council
WCMC	World Conservation Monitoring Centre
WTO	World Tourism Organization
WWF	World Wide Fund for Nature
ZICO	Zones importantes pour la conservation des oiseaux (Important zones for bird conservation)
ZNIEFF	Zones naturelles d'intérêt écologique, faunistique et floristique (Nature zones of ecological, faunal and floral interest)

List of Illustrations

List of Tables

List of Maps

Reference map. *(cf. Map 1–5)*

Summary

(1) Issues and Background

The Convention on Biological Diversity resulting from the United Nations Conference on Environment and Development (UNCED) in Rio de Janeiro in June, 1992 is the first international legal instrument to provide a comprehensive global basis for the various international efforts to conserve nature. Agreements developed in the past were either regional or limited by the scope of application. The Convention requires parties to conserve and, at the same time, sustainably use the biological diversity of the earth – understood as the diversity of genes, species and ecosystems.

Tourism is a way to use global biodiversity, and it may have impacts on it. Yet, Agenda 21 which makes the objectives of the Convention more precise, and translates them into recommendations for action, contains only marginal references to tourism. For quite some time, there has been a consensus supported by numerous studies that tourism has grave impacts on nature and the environment where it occurs in a massive form or when it uses particularly sensitive ecosystems.

As tourism activities concentrate on coastal areas the purpose of the present study is twofold: firstly, to analyse where and to which extent conflicts exist, or may be expected to arise, between conservation of biological diversity and development of tourism; secondly, to work out approaches and concepts which may be used to avoid and settle such conflicts. The global level is considered first; a discussion of the European context and case studies from various regions follow. The last part of the study concerns international-law aspects of sustainable tourism.

(2) Methodology

The terms of reference for the present study suggested a step-by-step concept consisting of the following elements:

1. Identification of sites of species and ecosystem types which are essential for global and European biodiversity;
2. Analysis of both the quantitative and qualitative features of coastal tourism and consideration of the areas on which it is presently concentrated;
3. Analysis of the existing potential of conflict between biodiversity and tourism and identification of most important causes, taking into account priorities with regard to areas and ecosystems;
4. Presentation of case studies of regional use conflicts as well as approaches to solve such conflicts in selected European coastal regions;
5. Description of concepts at the international, regional and national levels, including concepts of self-control to be applied by the tourism industry;
6. Analysis and assessment of existing rules in international law applicable to ensure sustainable tourism and recommendation for the further development of international law.

The findings in the technical part of the study are mainly based on a comprehensive assessment of available literature; in addition, statements and contributions made at the "World Conference on Sustainable Tourism" (Lanzarote, April, 1995) were taken into account. The results concerning points 1 to 3 above are presented in tables and maps, each with accompanying text.

(3) Biodiversity

Biodiversity is composed of the variety of genes, species and ecosystems. The central element is the species, as genes in general do not occur isolated in single individuals; they rather exist in combinations particular to each species. Of equal importance are ecosystems which provide the basis for the existence of species. Their destruction very often also threatens the existence of species which depend on them. On the other hand, the loss of species also threatens the existence of many ecosystems. The high value of biodiversity for humanity results from its functions to stabilize the biosphere, particularly the global climate, to provide a direct resource for consumption, and to support production. Meanwhile, the objective is globally recognised that biodiversity must be conserved. The basis of all conservation measures is a comprehensive inventory and assessment of biodiversity.

The present study provides an overview at both the global and the European levels and shows priority areas where a particularly high biodiversity exists requiring particular conservation measures.

(4) Global Tourism

With an average annual growth rate of 5.5%, tourism is among the fastest-growing activities worldwide. In 1993, more than 500 million international arrivals were counted (*WTO News* 1/1994) which, in economic terms, meant an income worth 324 billion US $. With an average rate of 12.6%, income grows significanty stronger than the numbers of arrivals. WTO estimates place tourism, together with petroleum and cars, among the leading branches in world trade. In 1990, tourism had a share of 12% of the global GNP, and 15% of the global turnover in services (WTO 1990). Meanwhile, these shares should have increased, according to the growth rates indicated above.

In Europe, the number of international arrivals increased from 190 million in 1980 to 288 million in 1992. The average annual growth rate is 3.5%. Thus, European tourism amounts to $^{2}/_{3}$ of global tourism. 35% of all international travellers per year gather in the Mediterranean alone.

(5) Impacts of Tourism

For a long time, tourism was considered a “white industry” with no need to point at its impacts on the social, cultural and ecological environment. First voices of criticism were raised in the 1970s. However, it took more than a decade to make people aware of the consequences of tourism (BMZ 1993).

Today, almost all actors in the field of tourism agree that tourism may have negative impacts on the environment and that those impacts should be controlled. However, Agenda 21 takes only little notice of the impacts of tourism on biodiversity.

The present study describes the major environmental impacts of tourism in coastal, mountain and natural areas. It then takes the example of coastal tourism to outline the ecosystem and area-based conflicts between biodiversity and tourism. The need and urgency of action at both the global and European levels to conserve biodiversity is being demonstrated.

(6) Case Studies

The impacts of coastal tourism on the biological diversity in the countries and the related conflicts of uses analysed at the global and European level are described in a more detailed way using case studies from European countries. The case studies refer to various European coastal regions (Côte d'Azur, Costa del Sol, Wadden Sea etc.) and deal with the different conflicts between conservation of biological diversity and touristic developments. At the same time the study presents approaches which have been developed in the respective countries and which aim at controlling the adverse and destructing impacts of tourism on coastal ecosystems and at conserving biodiversity.

(7) Strategies

Strategies to avoid and limit the impacts of tourism on nature and the environment are in principle available in national programmes, planning strategies, guidelines and recommendations; they have been developed by governmental, inter-governmental and non-governmental organisations as well as by the tourism industry itself. Despite that, the stress on and destruction of ecosystems continue. We observe at the global level a continuous loss of species and a loss of habitats which are intact, in the sense that they "function" from an ecological point of view.

The strategies developed so far are not very precise and have only little binding force, which is illustrated by the fact that there are many recommendations but no legally binding agreements addressing explicitly the relationship between biodiversity and tourism. Most advanced in this respect are national laws and regulations in a number of destination countries; however, often implementation and effective control measures are missing, particularly in developing countries.

It is basic to recognize, at both the international and the national level, that touristic uses are threatening conservation of biological diversity and that specific strategies need to be applied to avoid conflicts.

The present study provides an overview of existing instruments and programmes which need to be further developed with regard to tourism, ideally combined and consistently enforced. If this can be achieved an effective strategy will be available to conserve biological diversity and to make tourism sustainable.

(8) International Law Issues

The legal part of the study deals with the need to regulate tourism in order to make it environmentally friendly. It is based on the results of the technical part of the study describing the environmental problems of tourism. Existing international agreements relevant to tourism at the global and the regional levels alike are being analysed.

The review of existing law leads to the result that, despite the economic and ecological importance of tourism, only first initiatives to develop legally binding regulations have been started. Within the Alps Convention framework a protocol on tourism is presently being negotiated. With regard to Antarctica debates have been going on in recent years to develop a specific legal instrument under the umbrella of the Antarctic Environmental Protocol. Many international conventions at both the global and regional level, particularly those on nature conservation, are in principle applicable to tourism activities; however, they do not deal specifically with tourism. Therefore, the study clearly identifies the need for legal regulations.

Legal regulations are needed at both the global and regional levels. Global regulations, by necessity, need to be more general; they call for more detailed regulation through regional agreements. The regulatory options are: separate agreements on sustainable tourism; agreements additional to existing (nature conservation) conventions; additional provisions within existing (nature conservation) conventions.

The study concludes that agreements (protocols) additional to existing nature-conservation conventions are the preferable option. With this option, negotiations would be guided by considerations of nature conservation, as the relationship between tourism and nature conservation is obvious: Effective nature conservation and tourism depend on each other. Tourism relies on an intact nature; nature conservation which does not take into account that tourism may have impacts on nature proves to be helpless.

As to the global level, the proposal is made to develop a protocol additional to the Convention on Biological Diversity. Basic elements of such a protocol are drafted and presented here. As a global agreement it is confined to basic rules which necessarily need to be more general. They need to be acceptable to a variety of states and groups of states while containing, at the same time, somewhat effective requirements.

It is possible, and from a political point of view advisable, to prepare such global agreement through legally non-binding guidelines. These guidelines could be developed along the lines of the basic elements which are proposed here for the protocol additional to the Convention on Biological Diversity.

Zusammenfassung

(1) Fragestellung und Hintergrund

Die Konvention über die biologische Vielfalt, ein Ergebnis der Konferenz der Vereinten Nationen für Umwelt und Entwicklung (UNCED) im Juni 1992 in Rio de Janeiro, hat die verschiedenen internationalen Naturschutzbemühungen erstmals auf eine umfassende, globale Grundlage gestellt. Bisher hatte es lediglich regional oder inhaltlich begrenzte Naturschutzabkommen gegeben. In der Konvention verpflichten sich die Parteien, die biologische Vielfalt der Erde – unter der sowohl die genetische Vielfalt als auch die Vielfalt von Arten und Ökosystemen verstanden werden – zu schützen und gleichzeitig nachhaltig zu nutzen.

Eine der Nutzungen, die Auswirkungen auf die globale Biodiversität hat, ist der Tourismus. Dieser wird in der Konvention über die biologische Vielfalt nicht und in der Agenda 21, dem ebenfalls in Rio verabschiedeten, rechtlich nicht bindenden, Aktionsprogramm für eine nachhaltige Entwicklung nur am Rande erwähnt. Dabei ist seit langem unbestritten und in zahlreichen Untersuchungen belegt, daß Tourismus, wo er in massiver Form auftritt oder besonders empfindliche Ökosysteme in Anspruch nimmt, schwerwiegende Auswirkungen auf Natur und Umwelt hat.

Da sich Tourismus vor allem auf Meeresküsten konzentriert, war es einerseits Aufgabe der vorliegenden Studie zu untersuchen, an welchen Küsten und in welcher Intensität Konflikte zwischen dem Erhalt der biologischen Vielfalt und der touristischen Entwicklung bestehen bzw. zu erwarten sind. Zum anderen galt es, Ansätze und Konzepte zur Vermeidung und Lösung solcher Konflikte aufzuzeigen. Dazu sollte zunächst die globale Ebene betrachtet werden, im Folgeschritt die europäische und auf Länderebene Fallbeispiele aus verschiedenen Regionen. Der abschließende Teil der Studie befaßt sich mit internationalen rechtlichen Aspekten des umweltverträglichen Tourismus.

(2) Vorgehensweise

Den Fragestellungen des Untersuchungsauftrages entsprechend wurde das Vorhaben als mehrstufiges Konzept mit folgenden Untersuchungsabschnitten angelegt:

1. Identifizierung der für die globale und europäische Biodiversität wichtigsten Verbreitungsgebiete von Arten und Ökosystemtypen.
2. Analyse der quantitativen und qualitativen Merkmale des Küstentourismus und seiner aktuellen räumlichen Schwerpunkte.
3. Analyse des vorhandenen Konfliktpotentials zwischen Biodiversität und Tourismus, dabei Identifizierung der wichtigsten Auslösefaktoren sowie räumlicher und ökosystemarer Schwerpunktbereiche.
4. Darstellung von Fallbeispielen für regionale Nutzungskonflikte und Lösungsansätze in ausgewählten europäischen Küstenregionen.
5. Beschreibung von Lösungsansätzen auf internationaler, regionaler und nationaler Ebene sowie Ansätzen zur Selbststeuerung in der Tourismusindustrie.
6. Analyse und Bewertung vorhandener völkerrechtlicher Regelungen, die für die Gewährleistung eines umweltverträglichen Tourismus relevant sind und Vorschläge für die Weiterentwicklung des internationalen Rechts.

Die Aussagen beruhen vorwiegend auf den Ergebnissen einer umfassenden Literaturrecherche und wurden unterstützt durch Aussagen im fachlichen Teil der Studie sowie Fachbeiträgen von Teilnehmern der „Weltkonferenz für einen nachhaltigen Tourismus“, die im April 1995 auf Lanzarote stattfand. Die Bewertungsergebnisse der Pkt. 1–3 wurden tabellarisch und kartographisch dargestellt und mit entsprechenden Erläuterungstexten versehen.

(3) Biodiversität

Biodiversität setzt sich aus der Vielfalt von Genen, Arten und Ökosystemen zusammen. Zentrales Element ist die Art, da Gene im allgemeinen nicht isoliert in einzelnen Individuen vorkommen, sondern in artenspezifischen Kombinationen. Von gleich hoher Bedeutung sind Ökosysteme da sie die Grundlage für die Existenz der Arten darstellen. Ihre Zerstörung bedroht sehr häufig auch den Bestand der auf sie angewiesenen Arten. Umgekehrt gefährdet auch der Artenschwund die Existenz vieler Ökosysteme. Der Wert einer hohen Biodiversität für die Menschheit ergibt sich vor allem durch ihre Funktion als Stabilisator der Biosphäre (insbesondere des Weltklimas) und als direkte und produktionsunterstützende Ressource. Inzwischen gilt es weltweit als erklärtes Ziel, die Biodiversität zu erhalten. Grundlage für jede Schutzmaßnahme ist eine umfassende Bestandsaufnahme und Bewertung der Biodiversität.

Vorliegende Studie gibt einen Überblick auf globaler und europäischer Ebene und zeigt Schwerpunktbereiche auf, in denen eine besonders hohe Biodiversität existiert bzw. der Schutzbedarf besonders hoch ist.

(4) Welttourismus

Mit einer durchschnittlichen jährlichen Wachstumsrate von 5,5% zählt Tourismus weltweit zu den am stärksten wachsenden Aktivitäten. 1993 wurde im internationalen Tourismus die 500-Millionen-Marke an Ankünften erreicht (*WTO News* 1/1994). In ökonomischer Hinsicht bedeutete dies Einnahmen von über 324 Mrd. US $ (ebd.), welche mit jährlich durchschnittlich 12,6% deutlich stärker angewachsen sind als die Ankünfte. Damit gehört Tourismus nach Schätzungen der WTO im Welthandel zusammen mit Erdöl und Kraftfahrzeugen zu den führenden Wirtschaftszweigen. Tourismus hatte 1990 einen Anteil von 12% am globalen Bruttosozialprodukt und von 15% am globalen Umsatz im Dienstleistungssektor (WTO 1990). Inzwischen dürften sich diese Anteile entsprechend der Wachstumsraten noch beträchtlich erhöht haben.

In Europa ist die Anzahl der internationalen Ankünfte von 190 Mio. (1980) auf 288 Mio. (1992) angestiegen. Die durchschnittliche jährliche Wachstumsrate beträgt 3,5%. Europa trägt damit allein $^{2}/_{3}$ des Welttourismus (WTO). Allein im Mittelmeerraum konzentrieren sich jährlich 35% aller international Reisenden.

(5) Auswirkungen von Tourismus

Tourismus galt lange Zeit als „weiße Industrie“, Auswirkungen auf seine soziale, kulturelle und ökologische Umwelt blieben lange unangesprochen. Erste Kritik an den Folgen des Tourismus wurde bereits in der 70er Jahren geübt. Es dauerte jedoch mehr als ein weiteres Jahrzehnt, ehe diese Erkenntnis allgemeine Verbreitung und Akzeptanz fand (vgl. BMZ 1993).

Heute ist es unter den am Tourismus beteiligten Akteuren fast ein Allgemeinplatz, daß Tourismus negative Umweltauswirkungen haben kann, und daß diese so weit wie möglich begrenzt werden sollten. Dennoch wurden die Auswirkungen des Tourismus auf die Biodiversität in der Agenda 21 nur am Rande erwähnt.

Die Studie stellt die wichtigsten Umweltauswirkungen von Küsten-, Gebirgs- und Naturtourismus dar und zeigt in einem weiteren Schritt exemplarisch für den Küstentourismus die ökosystemaren und räumlichen Konfliktbereiche von Biodiversität und Tourismus auf. Handlungsbedarf und Dringlichkeit zum Erhalt der Biodiversität werden im weltweiten und europäischen Vergleich dargestellt.

(6) Fallbeispiele

Die auf globaler und europäischer Ebene betrachteten Auswirkungen des Küstentourismus auf die biologische Vielfalt in den Ländern und die damit verbundenen Nutzungskonflikte werden anhand von Fallbeispielen aus den europäischen Ländern vertiefend dargestellt. Die Fallbeispiele stellen anhand von verschiedenen europäischen Küstenregionen (Côte d'Azur, Costa del Sol, Wattküste, etc.) unterschiedliche Konfliktsituationen zwischen dem Erhalt der biologischen Vielfalt und der touristischen Entwicklung dar. Gleichzeitig werden Lösungsansätze aus diesen Ländern aufgezeigt, mit denen versucht wird, den Belastungen und Zerstörungen wertvoller Küstenökosysteme durch Tourismus entgegenzuwirken und die Biodiversität zu erhalten.

(7) Lösungsstrategien

Lösungsstrategien für die Vermeidung und Begrenzung der Auswirkungen des Tourismus auf Natur und Umwelt finden sich heute zumindest in Ansätzen in vielen Länderprogrammen, Planungsstrategien, Richtlinien und Empfehlungen sowie bei staatlichen, zwischenstaatlichen und Nicht-Regierungs-Organisationen und in der Tourismusbranche selbst. Dennoch schreitet die Belastung und Zerstörung der Ökosysteme kontinuierlich fort. Ein Rückgang der Artenvielfalt und ein Verlust an intakten, d.h. aus ökologischer Sicht „funktionierenden" Lebensräumen, ist weiterhin weltweit zu beobachten.

Die bisherigen Lösungsstrategien sind meist von geringem Konkretisierungs- und Verpflichtungsgrad, was sich u.a. darin niederschlägt, daß es eine Vielzahl von Empfehlungen, jedoch keine bindenden Abkommen gibt, die dezidiert das Verhältnis Biodiversität und Tourismus betreffen. Am weitesten gehen in dieser Hinsicht gesetzliche Regelungen in einer Reihe von Zielländern, jedoch fehlt es – insbesondere in Entwicklungsländern – sehr häufig an der Umsetzung bzw. effektiven Kontrollmöglichkeiten.

Grundvoraussetzung dafür ist, daß sowohl auf internationaler als auch auf nationaler Ebene anerkannt wird, daß touristische Nutzungen eine Gefährdung für den Erhalt der biologischen Vielfalt darstellen und es daher der Anwendung gezielter Lösungsstrategien zur Konfliktvermeidung bedarf.

Die Arbeit gibt einen Überblick über vorhandene Instrumente und Programme, die bei entsprechender Weiterentwicklung im Hinblick auf Tourismus, bei konsequenter Anwendung und Kombination zu einer wirkungsvollen Strategie für den Erhalt der biologischen Vielfalt an Küsten und einen nachhaltigen Tourismus genutzt werden können.

(8) Internationale rechtliche Fragen

Der rechtliche Schwerpunkt des Forschungsvorhabens befaßt sich mit der Regelungsbedürftigkeit des Tourismus unter dem Gesichtspunkt der Umweltverträglichkeit. Er stützt sich dabei auf die Ergebnisse des fachlichen Teils der Studie, in dem die Probleme des Tourismus unter dem Gesichtspunkt der Umweltverträglichkeit herausgearbeitet werden. Es werden bisher bestehende internationale Abkommen untersucht, die auf den Tourismus Anwendung finden können, und zwar auf globaler wie auch auf regionaler Ebene.

Es zeigt sich dabei, daß bisher, trotz der wirtschaftlichen und ökologischen Bedeutung des Tourismus, nur in ersten Ansätzen rechtliche Regeln im Entstehen begriffen sind. Im Rahmen der Alpenkonvention entsteht derzeit ein Protokoll über Tourismus; im Zusammenhang mit dem Naturschutzprotokoll für die Antarktis hat man in den letzten Jahren überlegt, ein spezielles rechtliches Instrument über Tourismus in der Antarktis zu schaffen. Es sind zwar eine Reihe internationaler – globaler und regionaler – Abkommen auf den Tourismus anwendbar, insbesondere Abkommen zum Naturschutz; die Regelungen sind allerdings nicht speziell auf die Problematik Umwelt und Tourismus zugeschnitten, so daß ein Regelungsbedarf gesehen wird.

Ein Regelungsbedarf wird dabei sowohl auf der globalen wie auch der regionalen Ebene gesehen. Globale Regelungen müssen notwendigerweise allgemeiner bleiben; sie bedürfen näherer Ausgestaltung durch regionale Vereinbarungen. Die Regelungsoptionen sind folgende: selbständige Abkommen über umweltverträglichen Tourismus; Zusatzabkommen zu bestehenden (Naturschutz-) Abkommen; Zusatzregelungen innerhalb bestehender (Naturschutz-) Abkommen.

Im Ergebnis wird in dieser Frage die Meinung vertreten, daß Zusatzvereinbarungen (Protokolle) zu bestehenden Naturschutzabkommen die vorzugswürdige Lösung darstellen. Inhaltlich ergäben sich bei dieser Lösung Orientierungspunkte für die zu schaffenden Vereinbarungen, denn die Zusammenhänge zwischen Naturschutz und Tourismus liegen auf der Hand: ein wirksamer Naturschutz und ein nachhaltiger Tourismus bedingen sich gegenseitig. Tourismus ist heute ohne eine intakte Natur nicht möglich; Naturschutz, der die möglichen Beeinträchtigungen der Natur durch touristische Aktivitäten nicht berücksichtigt, ist defizitär.

Auf globaler Ebene wird hier vorgeschlagen, ein Zusatzprotokoll zur Konvention über die biologische Vielfalt zu entwickeln. Ein solches Zusatzprotokoll wird in seinen Grundelementen im Text vorgestellt. Es beschränkt sich, da es ein weltweiter Vertrag sein muß, auf grundsätzliche Regelungen, die hinsichtlich ihrer Regelungsdichte von einer gewissen Abstraktion sein müssen. Sie müsen für eine Vielfalt von Staaten und Staatengruppierungen akzeptabel sein, dabei natürlich noch wirksame Anforderungen enthalten.

Es ist denkbar und rechtspolitisch empfehlenswert, solch ein weltweites Abkommen durch rechtlich nicht bindende Richtlinien vorzubereiten. Diese Richtlinien könnten sich inhaltlich schon an den Elementen orientieren, die für das Zusatzprotokoll ausgearbeitet und hier vorgestellt werden.

Section A. The Global Situation

The "Convention on Biological Diversity", adopted at the United Nations Conference on Environment and Development (UNCED) in June, 1992 in Rio de Janeiro, created for the first time a comprehensive global foundation for various international endeavours to protect nature. Up to that time, nature-conservation agreements had been limited to a region or objective. In the Convention the signatory states pledge to protect the earth's biological diversity – which includes genetic diversity as well as the diversity of species and ecosystems – and at the same time to make use of them in a sustainable way.

One of the uses having an impact on global biodiversity is tourism, but in Agenda 21, in which the goals of the Convention are more precisely defined and brought to bear in a series of recommended steps, tourism only features marginally, mostly in regard to mountain and coastal ecosystems. The task of this study is, firstly, to ascertain where and to what extent conflicts between nature-conservation goals and tourism exist on a global scale. Secondly, examples will be provided for minimising these conflicts. To this end, a four-step procedure will be followed:

1. identification of the species habitats and ecosystem types which are most important to global biodiversity;
2. analysis of the quantitative and qualitative characteristics of global tourism, particularly in regard to geographical distribution patterns and the impacted ecosystem types;
3. analysis of the impact of tourism on species and ecosystems, with special attention to identifying the most important impacting factors and major regions and ecosystems;
4. description of solution strategies at the international, regional and national levels as well as self-regulating approaches in the tourism industry.

1 Global Biodiversity

The World Conservation Monitoring Centre (WCMC), a body of the International Union for Conservation of Nature and Natural Resources (IUCN), has compiled a first survey and evaluation of biodiversity on a global scale ("Global Biodiversity – Status of the Earth's Living Resources"). Although there is still a considerable lack of knowledge in this regard in some parts of the world, the study provides a comprehensive overview and identifies major areas where a particularly high biodiversity exists or the need for conservation is especially urgent. The statements herein are thus mainly taken from this publication. However, it should be pointed out that the IUCN's rankings are purely quantitative data evaluations based mainly on species count. But the concept of biodiversity also includes qualitative aspects of individual ecosystems. Historically evolved, natural and culturally influenced landscapes as well as nature areas are not included in the IUCN's survey. Here further statistical surveys are needed which do justice to the qualitative aspects of ecosystems. A comprehensive survey on biodiversity is contained in the reference work of the UNEP entitled "Global Biodiversity Assessment", published in 1995. Here, a first attempt is made to analyse various ecosystem types and to correlate them with human use and its impacts on them. However, tourism as a special form of this use is not discussed, and all other impacts are described at a very general level. The United Nations Environmental Programs (UNEP) and the European Union (EU) are currently conducting a pilot study in ten bio-geographical regions of the world in which information on land use, habitat structure and landscape types are being assessed, along with data on animal and plant species, all with regard to biological and landscape diversity. The results are not yet available.

Consequentially, in this study the term "species diversity" is used for surveys based solely on species count, whereas for data reflecting the variety of ecosystems as well as species, the term "biodiversity" is used.

1.1 Basic Criteria

"Biological diversity describes life in all the fascinating variety it manifests itself in. As defined by the agreement, biological diversity means the variety of living organisms of any origin, including terrestrial, marine and other aquatic ecosystems as well as the ecological complexes which they are part of. This encompasses the diversity both within a given species and among species as well as the diversity of ecosystems" (KNAPP 1995, p. 7).

Biodiversity comprises – as stated above – the following elements:

- diversity of genes
- diversity of species
- diversity of ecosystems.

The central element is the species, as isolated genes generally do not occur in single individuals, but in combinations which define the species. Ecosystems are also highly significant, as they constitute the basis for the species' existence, and their destruction is very often a threat to the survival of the species depending on them; similarly, the extinction of species often jeopardises the existence of many ecosystems.

The value of high *biodiversity* for man may be sketched as follows:

1. *as a resource:*
   – directly (for the production of food and medicine)
   – to support production (e.g., forests for erosion protection and water storage, coral reefs for fish reproduction);
2. to stabilise the *biosphere* (above all global climate);
3. as an *immaterial* (ethical, cultural, aesthetic) value, which can be "transformed" into material values via tourism (WCMC 1992).

Species are highly significant for global biodiversity when characterised by the following:

- threat of extinction (Red List species)
- endemic, taxonomically isolated species (e.g., lemurs on Madagascar)
- relatives of domesticated/tamed species living in the wild
- high medicinal potential
- high social/cultural significance.

Of particular significance are the taxonomically isolated species, as they have little similarity to other species and are therefore unique in regard to genetic structure, too. These species are also often endemic to limited areas.

Their extinction amounts to a greater loss for global biodiversity than the extinction of a species which has a large number of related species.

Ecosystems are highly significant for global biodiversity when characterised by the following:

- high number of species
- habitat of endangered species
- habitat of endemic species
- important for migratory species
- especially pristine, unique or representative
- high social/cultural significance (GLOWKA et al. 1994, p. 34).

1.2 Regions of High Significance for Biodiversity

The highest species diversity in purely quantitative terms is found in tropical countries. More temperate climate zones are in contrast relatively insignificant. The same is true for areas with high precipitation in comparison with dry regions. The IUCN has identified 12 *"megadiversity"* countries, in which 70% of all vertebrates and higher plant species live: Mexico, Columbia, Ecuador, Peru, Brazil, Zaire, Madagascar, China, India, Malaysia, Indonesia and Australia (WCMC 1992). When the degree of endemicity and species density is also taken into account, a few other countries can be added, such as Costa Rica *(see Fig. 1)*.

Fig. 1. Countries of highest biodiversity. Source: WCMC 1994, p. 142, altered.

Regions having a high *degree of endemicity* have been the subject of various studies. Thus, endemic plants are very frequently found in the tropical rain forest and zones with climates similar to the Mediterranean (such as, for example, the Cape region in South Africa; cf. WCMC 1992). As to animal species, the degree of endemicity is particularly high in Australia, Southeast Asia, Madagascar and the Amazon region. Endemic bird and plant species are very frequently found on small oceanic islands (South Pacific, Caribbean, Indian Ocean).

The *Centers of Plant Diversity* (CPD) identified by the IUCN combine species diversity with degree of endemicity and endangerment. It may be assumed that these regions are not only highly significant for plant species, but for animal species, as well, as they usually combine a great number of various types of biotopes. Their size ranges from 5 (Jatún Sacha, Ecuador) to over 1 million sq. km. (Gran Chaco, South America). In the developing countries, the threat is usually posed by agriculture and forestry, as well as by mining. In the industrial countries (Europe, North America, Japan, Australia) and in the Caribbean and Oceania, tourism has become a major impacting factor.

Endangered species are also found primarily in tropical countries. The list is topped by Madagascar for mammals and Indonesia for birds. The greatest number of endangered reptiles, amphibians and fish species are in the United States. In this connection some remarks should be made. The great number of endangered species in tropical countries is primarily due to the great number of species found there. On the other hand, the real degree of endangerment could be much higher, as hardly anything is known of the existence of many species in the tropics. Relatively complete Red Lists are only available for developed countries.

Species extinction

According to estimations made by various authors, the global extinction rates of species are

- ca. 2,000 plant species per year in the tropics – a global loss per decade of 8 % (*Raven 1987*),
- ca. 25 % of all species between 1985 and 2015 – a global loss per decade of ca. 9 % (*Raven 1988*),
- at least 7 % of plant species (*Myers 1988*),
- 0.2 to 0.3 % of all species per year (*Wilson 1988*),
- 5 to 15 % of forest species by the year 2020 (*Reis and Miller 1989*),
- 2 to 8 % of all species between 1990 and 2015 (*Reid 1992*),
- 3 to 130 species daily (*Wiss. Beirat d. Bundesregierung, 1995*).

Source: Reid (1992) cit. ex WCMC (1992), Wiss. Beirat d. Bundesregierung Globale Umweltveränderung, WBGU 1995.

Fig. 2. Species extinction.

The principal threats to mammals and birds stem from *habitat destruction*, followed by hunting and the introduction of alien species *(cf. Fig. 3)*. In regard to habitat, most of the endangered mammalian and bird species are in the tropical rain forests, predominantly in low-lying regions. Steppes, freshwater systems, mountainous areas, deciduous forests and coastal ecosystems continue to be strongly impacted *(cf. Fig. 4)*. When reptiles, amphibians and fishes are taken into consideration, wetlands take on increased importance (WCMC 1992).

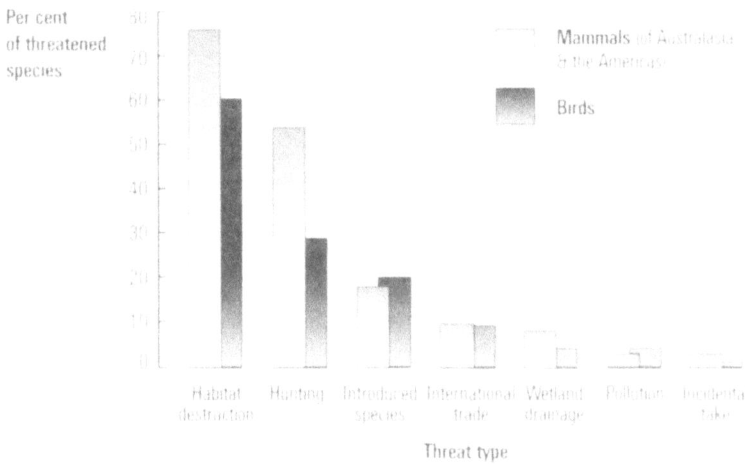

Fig. 3. Causes of danger to mammal and bird species. Source: WCMC 1992, Fig. 17.2, p. 236, altered.

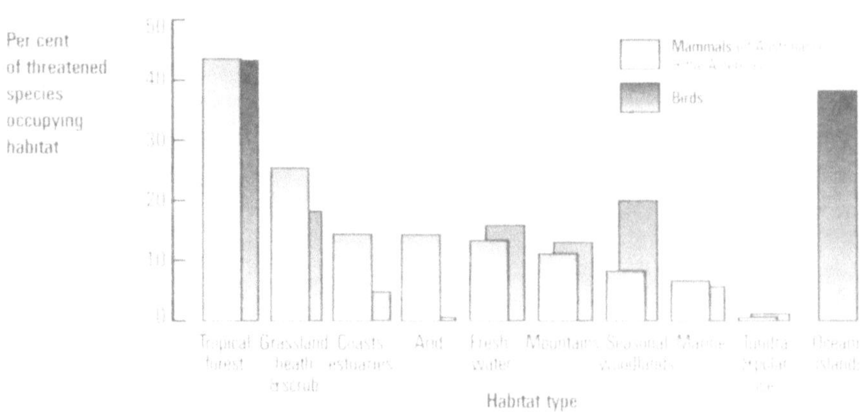

Fig. 4. Per cent of endangered species by habitat. Source: WCMC 1992, Fig. 17.6, p. 244, altered.

Particularly significant are *oceanic islands*, as they have a very high proportion of endangered endemic bird and plant species. Over 50% of the earth's endangered endemic bird species live on only 11 island groups or states: Hawaii, the Philippines, Indonesia, Papua-New Guinea, Solomon Islands (Melanesia), Marquesas Islands (Polynesia), New Zealand, Mauritius, Seychelles, São Tomé and Principé (West Africa) and Cuba. The impacted species live not so much on the coast as in the forests in the islands' interior and are endangered by the increasing destruction of these habitats (WCMC 1992).

Endangered endemic plants are found primarily on the following islands: Cuba (over 800 species!), Hawaii, Jamaica, Canary Islands, Mauritius, Galápagos, Socotra and Juan Fernández (Chile). Less extensively studied, but presumably with a very high share, are New Caledonia (Melanesia), Hispaniola, Taiwan and Fiji. The threat is posed primarily by introduced animal and plant species.

As to the threat to *ecosystems*, there is little detailed knowledge on a global scale. Assessments must therefore be limited to rough typifications. In general it is assumed that the following habitats are particularly severely endangered:

- freshwater ecosystems
- coastal ecosystems
- wetlands
- coral reefs
- ocean islands
- rain forests in temperate zones
- steppes in temperate zones
- tropical dry forests
- tropical rain forests
- mountain ecosystems

(GLOWKA et al. 1994, p. 41; BMU 1993, p. 101)

Of greatest significance by far for global biodiversity are the *tropical rain forests*. Their global distribution is shown in *Fig. 5*. Data on the annual loss of tropical rain forests are contradictory. This is also true of regions and individual countries. It may definitely be assumed that the greatest absolute losses are incurred in Brazil, the country with the largest share of rain forests. Among the countries with the highest deforestation rates is Costa Rica (WCMC 1992, p. 257f.). There are various causes for this, but it should be noted that tourism has virtually nothing to do with it (op. cit., p. 264f.). On the contrary, nature tourism is even expected to possibly contribute to the preservation of rain forests by discouraging other non-sustainable uses (cf. AGÖT 1995).

Map 1. Biodiversity in coastal regions.

1 Ranking in agreement with ELDER/PERNETTA 1991.

2 Source: WCMC 1994, Table 6.

3 Source: WCMC 1992, p. 244–247.

4 Source: DUGAN 1993; WCMC 1992, Fig. 22.2.

The establishment of *protected areas* is seen as the best possibility for preserving particularly valuable species and ecosystems (in-situ protection). Further options are hunting and gathering restrictions, banning the sale of endangered species etc. Worldwide 7.7 million sq. km. are under natural protection (a total of 8,491 protected areas; op. cit., *Table 18)*. This amounts to about 5.2% of the earth's land area (WCMC 1992, p. 451f.). However, country by country and in regard to the global bioregions, the figures are at times widely divergent. Whereas, for example, in Thailand, Indonesia, Tanzania or Venezuela over 10% of the land area is under protection, the corresponding figures in other countries which are also significant as regards biodiversity, such as Brazil, Zaire or Papua-New Guinea, are markedly lower (op. cit., p. 460f.).

As to bioregions, subtropical or temperate rain forests, mountain systems, island systems and tundras are represented far above average in the protected areas, whereas lake systems and steppes in temperate zones are underrepresented. Tropical rain forests and savannahs, which are important to global biodiversity, are slightly below average (op. cit., p. 452).

On a global scale, the largest share in area among the protected areas is taken up by the national parks (category II, IUCN); the largest number of protected areas is in category IV, Managed Nature Reserves (see chapter 6.3). The figures are only averages and vary from region to region. In Europe, for example, where comparatively few "natural" habitats remain, "protected landscapes" (category V, IUCN) are predominant, whereas in Australia, which has vast untouched landscapes, the national-park category makes up the largest share (cf. WCMC 1992, pp. 447–451).

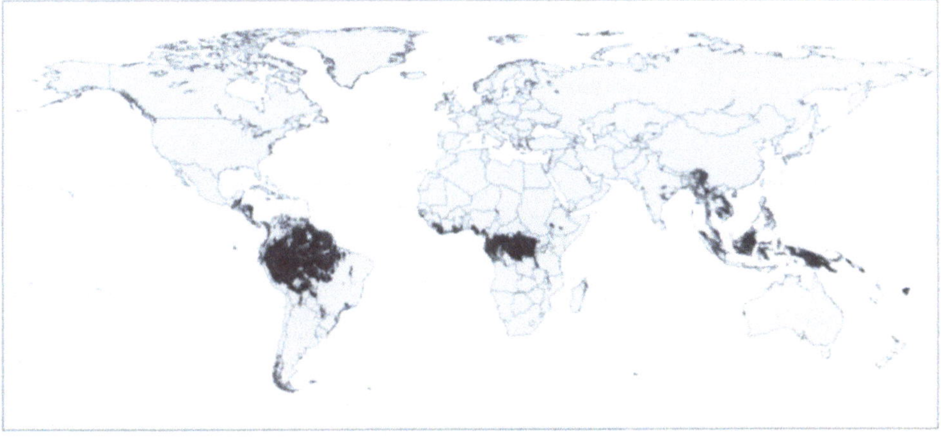

Fig. 5. Global distribution of tropical rain forests. Source: WCMC 1994, Fig. 6, p. 146, altered.

1.3 Coastal Ecosystems

As tourism is concentrated primarily on seacoasts (see chapter 2.2), such ecosystem types will be considered herein in more detail. Of the worldwide most severely endangered habitats named above, *wetlands*, *coral reefs* and *oceanic islands* fall into this category. The coastal areas having the highest species diversity are in Southeast Asia and the South Pacific *(cf. Table 1)*.

Species diversity in coastal regions	
Arctic Ocean	low
North Atlantic	low
Mediterranean Sea	low
Caribbean Sea	medium/low
Southwest Atlantic (South America)	low
Southeast Atlantic (West Africa)	medium
Western Indian Ocean (East Africa and Arabian Sea)	medium
Southeast Asia	high
South Pacific	high
North Pacific	low
Southeast Pacific (South America)	medium
Antarctic Ocean	low

Table 1. Species diversity in coastal regions. Source: ELDER/PERNETTA 1991. *(Authors' note:* The high species diversity in Southeast Asia and the South Pacific is primarily due to the many islands in these regions, most of them with a multitude of ecosystem types in a confined space; cf. p. 23.)

In general, seacoasts can subdivided into four *zones:*

- sea
- transitional area (often beaches)
- land area close to sea (e.g., dunes)
- hinterland (cf. PEARCE/KIRK 1986 , Fig. 1).

These zones are in turn taken up by various *ecosystem types (cf. Table 2).* The particular regional features of the historically evolved, natural and culturally influenced landscape structures and nature areas are reflected here most markedly. In their uniqueness they are of crucial significance for biological diversity.

Coastal zones and ecosystem types

Zone	*Ecosystem types*
Sea	open sea
	sea near coasts
	seafloor
	sand- and mud banks
	wadden sea (Europe)
	coral reefs (only in the tropics)
	lagoons of atolls and barrier reefs (only in tropics)
	eelgrass meadows
Transitional zone	sandy beaches
	pebble beaches
	lagoons of brackish water
	river mouths (deltas, estuaries)
	mangroves (only in the tropics)
	rocky coasts/cliffs
	clay cliffs
	small rocky islands
Land zone near the sea	coral islands (atolls, cays) (only in the tropics)
	dunes
	salt meadows
	brackish swamp forests
	freshwater wetlands (marshes, swamps, lakes)
Hinterland	various land ecosystems

Table 2. Coastal zones and ecosystem types. Source: compilation based on EUCC 1994, CORINE Biotopes, ESCAP 1992, WCMC 1992, ELDER/PERNETTA 1991, TARNAS/HASSAN 1990, DUGAN 1993.

Coral reefs are among the most productive ecosystems on earth and among the richest in species diversity. In this regard they are often referred to as the "rain forests of the ocean". Their global distribution is shown in *Fig. 6*. As they can only exist between the 30th parallels (north and south), 60% of them are in the Indian Ocean, 25% in the Pacific and 14% in the Caribbean. Coral reefs grow in shallow waters up to a maximum depth of 100 metres. A distinction should be made between barrier reefs (on the edge of mainland shelves), atolls (in the open ocean on extinct volcanoes) and seam and shelf reefs near coasts. Growing reefs also often form coral islands (Marshall Islands, Maldives, Florida Keys; WILKINSON/ BUDDEMEIER 1994, p. 10).

Apart from their high species diversity, coral reefs also fulfill important coastal-protective functions (as natural breakwaters) and for the reproduction of fish and crustaceans. They depend on high amounts of light, i.e., clear water, and on predominantly oligotrophic nutrient conditions. For these reasons they are very sensitive to sedimentation, sediment turbulence and eutrophication. They are also very vulnerable to mechanical damage (e.g., from storms or boats' anchors). In this regard, reefs near the coasts of countries under intensive industrial and agricultural exploitation are the most severely endangered (pollution and sedimentation from sewage and soil erosion; op. cit., p. 13f.). In many cases tourism is one of the contributing causes of reef deterioration (WCMC 1992, p. 307f.).

Wetlands are for the most part found in the Northern Hemisphere (Canada, U.S., Russia) and in the continental tropics (South America,

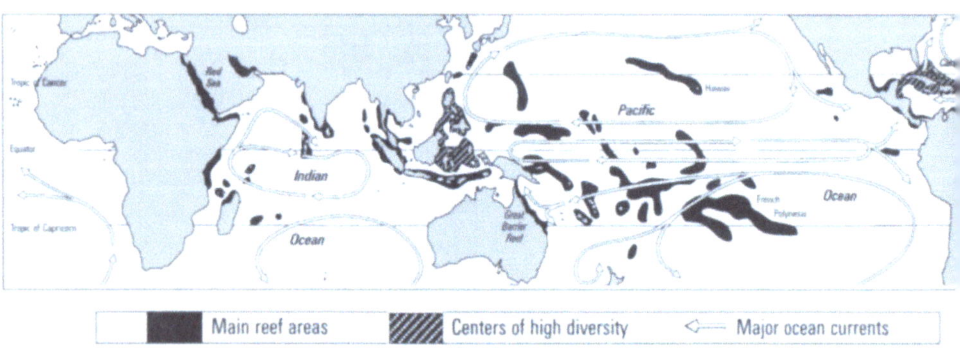

Fig. 6. Global distribution of coral reefs. Source: UNEP 1995, p. 382, altered.

Africa), predominantly in the interior. On the coast there are larger areas in the following regions: North America, Gulf of Mexico and the Caribbean, northeastern South America, southern Chile, Russia, the North Sea, West Africa, Southeast Asia and Australia. The most severe causes of stress on the wetlands are drainage and landfill measures in connection with construction projects (op. cit., pp. 293ff.).

Among the most significant tropical wet ecosystems are *mangrove forests*. Mangroves are to be sure less species-rich than coral reefs, but they are also very productive and assume an important function for coastal protection and fish and crustacean reproduction. Their geographical distribution is similar to that of coral reefs, but they grow in the transitional area between land and sea on shallow, muddy, nutrient-rich coasts, particularly at river mouths. Between mangroves and coral reefs there is at times an almost symbiotic relationship: whereas mangroves prevent nutrients and sediments from reaching the corals growing in deeper waters, the latter protect the mangrove forests from erosion by surf from the open sea (WILKINSON/BUDDEMEIER 1994, p. 18).

The countries with the greatest mangrove forests (over 1 million hectares each) are Indonesia, Nigeria, Mexico and Australia. Apart from general urbanisation and industrialisation, tourism is also one of the major causes of stress on them. In contrast to the coral reefs, eutrophication and sedimentation are not major factors, but rather construction projects involving direct destruction and subsequent landfills. As the major mangrove areas – shallow coasts and, particularly, river mouths – are also favourable sites for human settlement, this ecosystem is particularly impacted by damaging exploitation (WCMC 1992, p. 324f.).

Oceanic islands in the tropics often unite all the ecosystem types mentioned above within a confined space. In addition, there is a high proportion of endemic bird and plant species. They are therefore highly significant for global biodiversity. Particularly on smaller islands with limited resources (e.g., drinking water), environmental problems can quickly become severe if nothing is done to combat them. This is also true of tourism, which for many island states is the most important source of income today.

The existence of *protected areas* on the coasts is, first of all, an indication that particularly valuable ecosystems exist there, but also a reaction to a potential threat. Due to their sensitivity, protected areas are, on the one hand, more vulnerable to disturbances; but on the other hand they are better protected from them. In the nature-tourism sector, protected areas (marine and terrestrial) are becoming increasingly attractive. For this reason more and more conflicts should be expected; but then the danger of massive impairment due to tourism will be less than in likewise attractive but unprotected ecosystems. Marine and coastal protected areas thus constitute zones in which tourism and the protection of biodiversity will increasingly confront one another. *Figure 7* shows the global distribution of coastal and marine protected areas, thus illustrating the major regions of existent and/or expected "nature tourism". All the types of protected areas according to IUCN criteria are included in the figure.

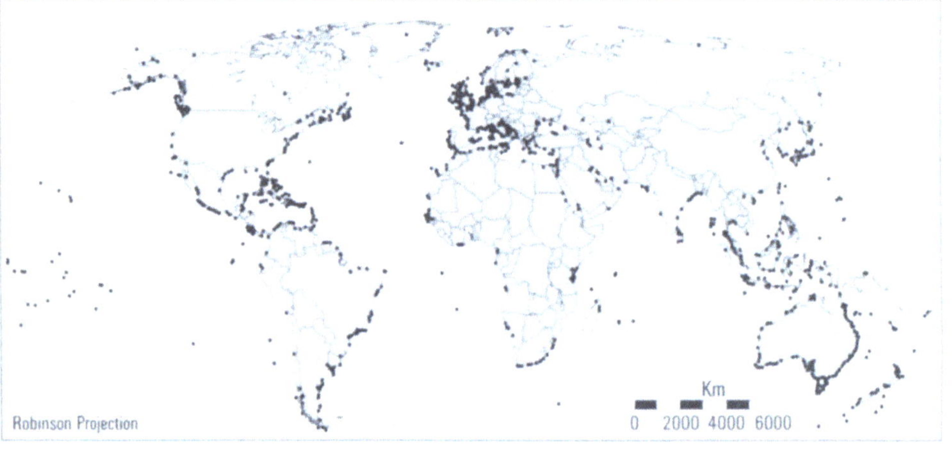

Fig. 7. Global distribution of marine and coastal reserves. Source: WCMC 1992, Fig. 29.6, p. 457, altered.

2 The Development of Tourism

"Tourism" is here defined as "the activity of people who travel to places outside their customary surroundings and stay there for leisure-time, business or other purposes for no longer than one year without interruption" (WTO, n.d.). Moreover, "their main reason for travel must be other than performing an activity which is remunerated at the place visited" (op. cit.).

A differentiation as to the various reasons for travel (holiday and recreational, business, educational travel etc.) cannot be made due to the widely varying statistical data in the various countries. There is no way of comparing the data: the World Tourism Organization (WTO) has indeed attempted to arrive at unification, but its annual statistics have been largely limited to the number of international arrivals in the destination countries. The average length of stay of visitors from abroad is recorded systematically only in some of the countries. As to reasons for travel, distinction is only made between business, holiday and "other" (visiting relatives/friends, spa/health vacation, pilgrimage), but even these data are not available for all countries. Completely missing is domestic tourism, which experts estimate at about ten times the volume of international travel (BENTLEY 1993). The figures quoted here therefore denote *international arrivals*.

2.1 Quantitative Development

2.1.1 Global Development

An exact quantification of the masses of worldwide tourism is, as stated above, hardly possible.

Despite these limitations, there is agreement that tourism is among the fastest-growing activities in the world, with average annual growth rates of 5.5% in the past 10 years (WTO 1994a). In 1993, international tourism topped the 500-million level (WTO News 1/1994). In economic terms, this meant revenues of 324 thousand million US $ (op. cit.), which, at an annual average of 12.6%, have grown markedly faster than arrivals (WTO 1994a).

According to WTO estimates, tourism ranks alongside petroleum and motor vehicles among the leading branches of world commerce. In 1990 it amounted to 12% of the gross global product and 15% of global turnover in the service sector (WTO 1990). These figures have probably significantly increased in the meantime. Some experts are convinced that even now tourism generates greater gross revenues than any other business activity in the world, with 5 thousand million estimated arrivals (including domestic tourism).

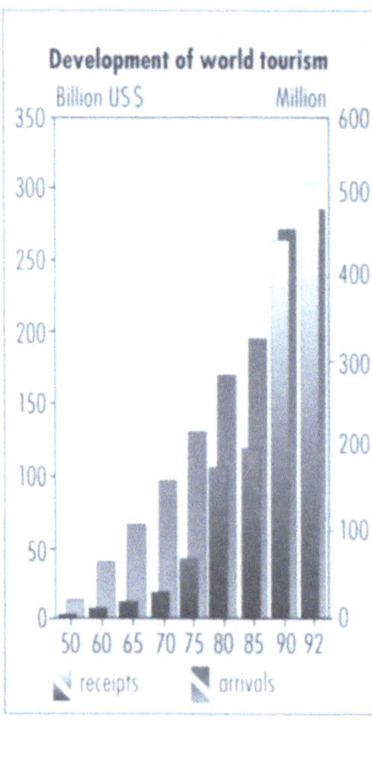

Fig. 8. Growth of global tourism. Source: WTO 1994a, p. 188, altered.

WTO prognoses do predict a slight decrease in growth in the coming years, but then international-tourism figures would nearly double once again by the year 2010, reaching 937 million arrivals (*WTO News* 1/1994). In addition, the share of long-distance and thus air travel will increase at an above-average rate of 6 to 7% annually (BENTLEY 1993; WTO 1990).

	Arrivals 1993 (million)	Share of international tourism (%)	Increase since 1988 (%)	Annual rate of increase, 1990–2000 and 2000–2010
Europe	296.5	59.3	18.0	2.5
America	106.5	21.3	27.8	4.6 / 3.5
East Asia/Pacific	68.5	13.7	51.9	6.8 / 6.5
Africa	17.9	3.6	43.3	5.0 / 4.0
Middle East	7.2	1.4	3.4	?
South Asia	3.4	0.7	18.0	6.1 / 6.0

Table 3. Regional development of tourism. Source: WTO 1994a, p. 190/191; *WTO News* 1/1994, 2/1994 and 5/1994, altered.

2.1.2 Regional Development

Europe accounts for nearly two thirds of global tourism, but this share has been decreasing for years, particularly in favour of the East Asian and Pacific regions and, in part, Africa, as well *(see Table 3)*. Whereas a further slow-down in growth is expected in Europe, East Asia and the Pacific will continue to be the region of most dynamic growth in the world. In general demand appears to be saturated in the highly developed tourism areas, because this demand will be increasingly attracted to more "exotic" destinations.

Within the world's regions, however, tourism's distribution is extremely uneven in some places. The WTO has therefore subdivided them into *subregions*. In these subregions the following developments took place between 1988 and 1992 *(cf. Table 4)*:

1. The *economic* significance of tourism as regards income from foreign exchange is highest in the Caribbean and the island states of the Pacific, followed by the eastern Mediterranean, Southern Europe and Central America.
2. As to the *area intensity* of tourism (international arrivals per sq. km.), the ranking is similar: most intensive is tourism in Micronesia; Western Europe is in second place, followed by Southern Europe, the Caribbean and Polynesia. However, these figures reflect the actual burden only in the case of smaller island regions. In states of larger land area, local concentrations (particularly in coastal regions) can yield much higher figures (Australia, for example).
3. Percentagewise, the *fastest-growing* subregions are South Africa, Central America, the eastern Mediterranean, Micronesia and Southeast Asia. Noteworthy is the slight decrease in tourist statistics in Southern Europe. The WTO moreover predicts above-average growth rates for the subregions Eastern Europe and North Africa. It is assumed that Southern Europe will lose a further share of its usual demand to more distant subregions, particularly Africa and the Caribbean *(WTO News* 5/1994).

	Arrivals 1992 (million)	Share of global tourism (%)	Rate of increase since 1988 (%)	Arrivals per sq. km. (1992)[1]	Ø Average share of foreign-currency receipts (%)[2]
Western Europe	117.018	28.5	24.3	106.0	9.1
Southern Europa	86.339	17.9	-4.0	69.9	19.5
North America	76.659	15.9	20.3	3.6	10.9
East./Centr. Europe	49.118	10.2	30.4	2.1	9.4
Northeast Asia	34.025	7.1	31.6	3.4	3.6 (excl. Macao)
Northern Europe	27.702	5.8	17.9	17.6	6.5
Southeast Asia	21.498	4.5	46.6	5.8	8.2
Caribbean	11.655	2.4	14.4	49.3	51.8
South America	10.423	2.2	30.7	0.6	9.5
East. Mediterran.	10.042	2.1	67.5	12.4	19.6 (only Turkey)
North Africa	9.067	1.9	24.0	1.6	13.5
Middle East	7.021	1.5	13.8	1.1	8.7 (excl. Egypt)
South Africa	3.742	0.8	171.4	1.4	5.3
Australia	3.659	0.8	17.5	0.5	9.1
South Asia	3.509	0.7	21.8	0.5	16.7
East Africa	2.941	0.6	26.0	0.6	9.2
Central America	2.400	0.5	68.3	4.7	19.1
West Africa	1.403	0.3	17.0	0.2	12.8
Micronesia	1.402	0.3	65.5	498.0	?
Melanesia	0.453	0.1	34.4	0.8	23.2
Central Africa	0.318	0.1	20.5	0.0	1.3
Polynesia	0.269	0.1	-3.2	47.7	60.4 (excl. Am. Samoa)

Remarks:
[1] Countries only included for which touristic data were available.
[2] Average value of percent share of individual countries was computed, if available; extremely divergent values were left out.

Table 4. Touristic development and significance in subregions. Source: calculations based on data in WTO 1994a.

2.1.3 Developments in Individual Countries

Maps 2 (international arrivals) and *3* (tourism intensity) illustrate the quantitative volume of tourism in the individual countries.

Table 5 lists the ten major countries of global tourism. With the exception of Mexico and China these are all industrial countries in Europe and North America. They are followed by a number of countries from East and Southeast Asia, with a third group comprising countries of Latin America and North Africa, as well as Australia. Hardly developed touristically are most countries in Africa.

	International arrivals 1993 (million)
France	61
USA	46
Spain	41
Italy	26
Hungary	23
Great Britain	19
Austria	18
China	18
Mexico	17
Germany	15

Table 5. The most important destination countries in global tourism. Source: *WTO News* 3/1994, altered.

The picture undergoes radical change in some cases when *tourism intensity* is compared to the land areas of the various countries. In this perspective, many of the island states of the Caribbean, Micronesia and the Indian Ocean are the most heavily frequented by tourists, apart from the city-states. Among the medium-sized states, those in Europe are most intensively frequented touristically. In large countries (with areas exceeding 500,000 sq. km.) the corresponding figures are of course much lower. However, as tourism there is concentrated in relatively few areas, particularly on the coasts, there is considerable tourism intensity per unit of area in some places *(cf. Map 3; Table 6)*.

All countries	Arrivals (1992) per sq. km. of land area	**Countries with large land area (>500,000 sq. km)**	Arrivals (1992) per sq. km. of land area
Saint Maarten (Netherl. Antilles)	13,878	**France**	108.9
Bermuda	7,076	**Spain**	78.5
Malta	3,171	**Thailand**	10.0
Aruba (Netherl. Antilles)	2,808	**Mexico**	8.8
Saba (Netherl. Antilles)[1]	1,923	**Turkey**	8.4
Guam (Micronesia)	1,621	**Morocco**	6.2
St. Eustatius (Netherl. Antilles)	1,429	**U.S.A.**	4.8
U.S. Virgin Islands (Caribbean)	1,424	**Egypt**	2.9
Northern Mariana Islands (Micronesia)[1]	1,059	**Republic of South Africa**	2.4
Cayman Islands (Caribbean)	931	**Kenya**	2.0

[1] including daytime

© 1996 by and Felix Dahms (graphic)

Table 6. Countries/islands with the highest tourism intensity per unit of area (excluding city-states). Source: calculations based on WTO 1994a.

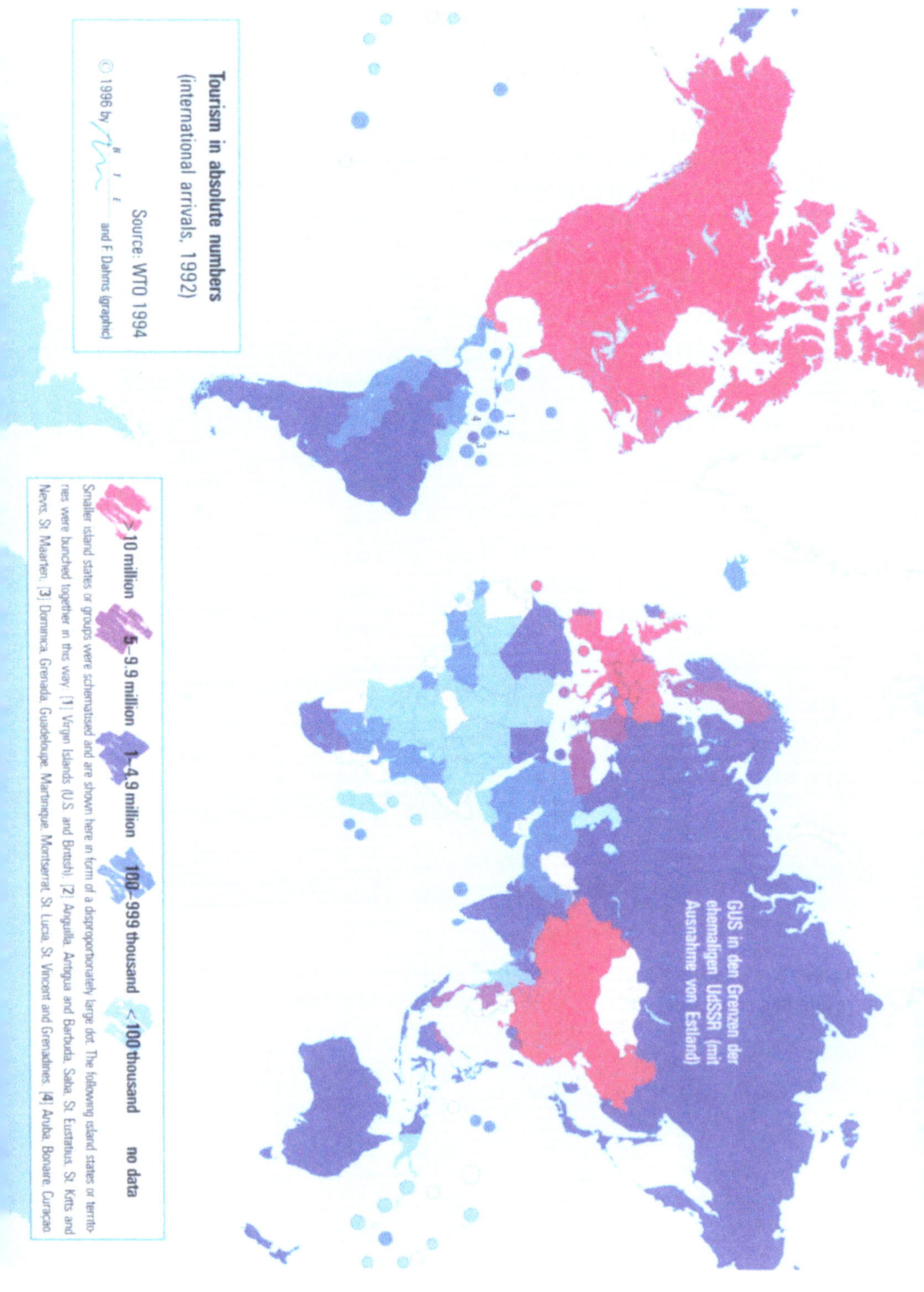

Map 2. Tourism in absolute numbers (international arrivals, 1992).

For a forward-looking estimate of global tourism, the *growth rates* in the various countries continue to be relevant. For practical reasons (comparing the WTO data over a longer period is often not feasible), the development from 1988 to 1992 was used here as reference base. This has the advantage that newer trends can potentially be included, but it also entails the risk that the figures can be distorted by unusual events at certain points or a re-organisation of national statistics (this may be why Gabun tops the list, for example). In contrast, the very high growth rate of the Republic of South Africa is plausible (because of the changed political situation). Indonesia continues to be of interest, as a high absolute level has already been reached there and this country is also of major significance for global biodiversity *(cf. Table 7)*.

In other highly developed tourist countries (with over 5 million international arrivals per year), the growth rates are much lower on average. This is also in agreement with WTO estimates, according to which the tourism centres of Europe – and parts of the Caribbean – have already peaked in their (quantitative) development or even exceeded it in some cases (negative growth rates in Italy, for example, the Bahamas, Barbados or the Virgin Islands). Nevertheless, very high growth rates are reported by some in this group, too, particularly by China.

All Countries	Increase in arrivals from 1988 to 1992 (%)	Leading tourist countries (> 5 million arrivals per year)	Increase in arrivals from 1988 to 1992 (%)
Gabun (Central Africa)	540	China	115
Rep. South Africa	386	Hungary	91
Iran	216	Turkey	76
Nicaragua	204	Malaysia	66
Vanuatu (Melanesia)	169	Denmark	47
Comoro Islands (Ind. Ocean)	138	Singapore	42
Indonesia	136	France	39
El Salvador	134	Portugal	34
China	117	U.S.A.	31
Sri Lanka	115	Hong Kong	25

[1] *increase from 1988 to 1991*

Table 7. Countries with the highest tourism growth rates. Source: calculations based on WTO 1994a.

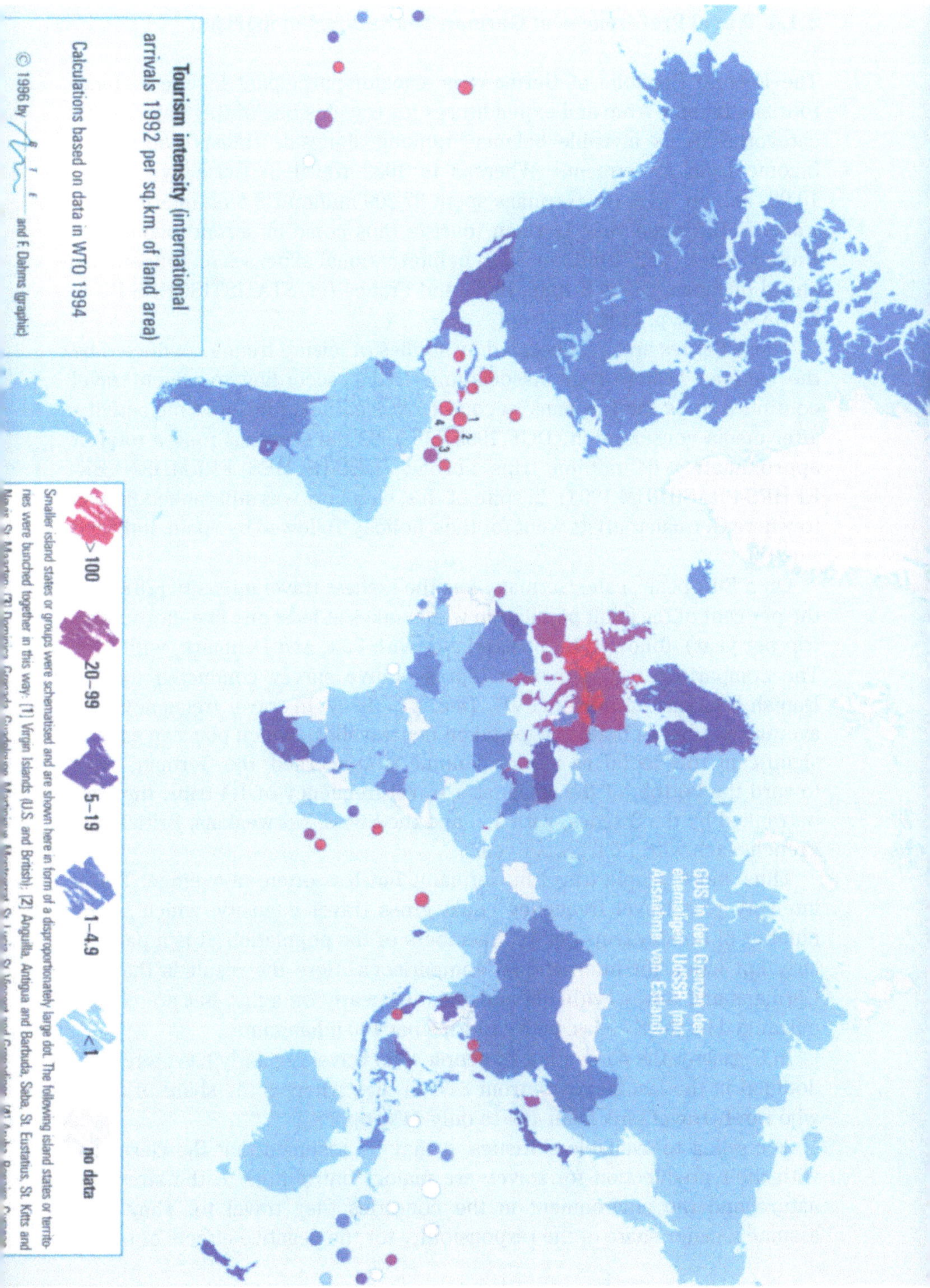

Map 3. Tourism intensity (international arrivals 1992 per sq. km. of land area).

2.1.4 Travel Preferences of German Tourists: a Comparison

The Federal Republic of Germany is a major participant in international tourism. Income from and expenditures for travel is one of the major single categories in its invisible balance, ranking alongside transportation and income from investments. Whereas in 1992 travel in Germany grossed 10,982 million US $, the Germans spent 37,309 million US $ on international travel in the same year. German tourists thus come in second behind the United States (39,872 million US $) in international expenses on tourism, far ahead of Japan, Great Britain, India and France (cf. STATISTISCHES BUNDESAMT 1995, p. 180f.).

These figures are corroborated by studies of leisure trends conducted by the German Leisure-Time Association (DGF), according to which travel continues to be the Germans' second most significant leisure-time activity, after media consumption (DGF 1995). In 1992 the Germans took a total of approximately 64 million trips abroad (DEUTSCHES FREMDENVERKEHRS-PRÄSIDIUM 1994). In spite of this, Germany was still ranked first as to where German tourists went for their holiday, followed by Spain, Italy and Austria.

On a European scale, Germany has the highest travel intensity (78%, i.e., the per cent of the adult population which takes at least one five-day holiday trip per year), followed by Switzerland, with 72%, and Denmark, with 71%. The comparison is based on a representative survey conducted for the Danish Tourism Authority (FVW, Dec. 27, 1995). In travel frequency (the average number of holiday trips taken per traveller), which plays an equally significant role as far as market volume is concerned, the Germans rank toward the bottom of the scale: at a travel frequency of 1.4 trips, they are outranked by the Swedes, with 1.7, and the Swiss, Norwegians, British and French, each with 1.6 trips per year.

Thus, more people travel in Germany, but less often on average. Travel intensity plus travel frequency equal gross travel intensity, which is the number of holiday trips per 100 members of the population. It is a particularly apt index for international comparisons. Here the result is that the Germans and Swiss, with 109 and 108, are nearly on a par, but are clearly outranked by the Swedes, with 116 trips per 100 inhabitants.

In Germany, the percentage of people who travel regularly has more than doubled in the last 20 years, from 24% to 51%, whereas the share of those who rarely travel sank from 44% to only 14% (op. cit.).

According to available statistics, it may be assumed that the Germans, with their predilection for travel, are major contributors to the stress on nature and the environment in the countries they travel to. They thus assume a major share of the responsibility for the negative effects of travel.

2.2 Qualitative Characteristics

Tourism can be classified according to the general categories holiday, business and other *reasons for travel* (visiting relatives/friends, spa/health vacation, pilgrimage). The exact proportion of each of these classes varies from country to country, but in general it may be stated that holiday travel constitutes the major share of global travel (WTO 1994a). The holiday-travel category in turn varies widely as to reason for travelling, activities undertaken, type of accommodations etc. A study of the travel preferences of German tourists, particularly in Third World countries, distinguishes between the following types of *holiday travel:*

- recreational holiday
- bathing/beach/sun holiday
- sports travel
- educational travel
- study/sight-seeing travel
- adventure travel
- pleasure travel

(BMZ 1993, p. 160f.)

However, these types of travel often overlap. For this reason, and because the result varies according to destination (domestic, European, long-distance travel), it becomes extremely difficult to make precise quantitative classifications. All in all, the categories recreational, bathing/beach/sun and pleasure travel clearly predominate. All other types of holiday travel are insignificant by comparison, with the exception of study and cultural travel and long-distance adventure travel (op. cit.).

A clear interpretation of these results as to the *nature* of tourist *destinations* is not possible, with the exception of beach holidays. Despite the lack of global data and the growing tendency of demand toward specialisation, it may nonetheless be assumed that the overwhelming share of international tourism is concentrated on the seacoasts (cf. BENTLEY 1993 etc.). A significant share of tourism continues to take place in cities, in the form of city sight-seeing, cultural and pleasure tourism. Moreover, mountains have a strong touristic appeal, as do, most recently, nature areas (cf. AGÖT 1995).

This study will be limited to those forms of tourism which impact ecosystems important to biodiversity. These are primarily:

1. *coastal tourism*
2. *mountain tourism*
3. *nature tourism*

City tourism is in this context of minor significance, as cities have generally already become insignificant for global biodiversity due to their "civilisational" overdevelopment. Moreover, in urban areas the effects of tourism are overshadowed by a number of other factors.

2.2.1 Coastal Tourism

Coastal tourism is by far the predominant form of tourism in the following regions:

- Mediterranean
- Caribbean and Gulf of Mexico
- island groups in the Indian Ocean
- Australia
- island groups in the South Pacific.

Intensively frequented are also the coasts of Western Europe, the United States, South and East Asia and individual points on the coasts of South America (particularly in Brazil, Uruguay and Ecuador) and Africa (the Maghreb countries, Egypt, West Africa, Kenya, Republic of South Africa).

Coastal tourism consists of a number of different activities, of which bathing and beach activities predominate by far. In addition, or as special forms of travel, the following *types of activity* play an important role:

- non-motorised aquatic sports (surfing, sailing)
- skin diving (particularly in coral reefs)
- motorised aquatic sports (motor boating, water skiing, jetskis, parasailing)
- angling and clam diving
- nature observation.

In addition, activies in the immediate coastal vicinity or in the hinterland, which are as such independent of the sea and beach, fulfill an important *complementary* function:

- swimming in pools
- sunbathing and picknicking (e.g., on dunes)
- hiking and bicycle trips
- sports (e.g., tennis, horseback riding, golf, motor sports)
- urban and social activities (shopping, bars, cultural events)
- sight-seeing excursions
- visits to nature areas.

(cf. EUCC 1994; ESCAP 1992; etc.)

As to *type of accommodations*, the following forms of tourism may be distinguished:

- mass tourism on urbanised coasts with high-rise buildings (particularly in Southern and Western Europe and parts of North and South America and Southeast Asia);
- resorts or holiday clubs in secluded, self-sufficient settings (frequently in developing countries such as the Maldives, Kenya, Dominican Republic);
- camping holidays (particularly in Europe and North America);
- nature tourism with sparse, relatively simple accommodations (e.g., the northwest coast of North America, Costa Rica, Belize, Australia).

All the forms of tourism named thus far originate from the land. In addition there is *cruise tourism* (on large passenger ships or small yachts), which is a factor particularly in the Caribbean and Mediterranean, but also, increasingly, in the Antarctic.

2.2.2 Mountain and Nature Tourism

Mountain tourism takes place on a particularly intensive scale in the Alps, but also at certain points in the Himalayans (India, Nepal), the mountains of East and Southeast Asia (Japan, China, Thailand), North America, Australia and New Zealand and the Andes (Ecuador, Peru, Chile, Argentina). The following forms of mountain tourism should be distinguished:

- nature and adventure tourism (hiking, trekking, rock and mountain climbing)
- winter sports (particularly alpine skiing, along with cross-country skiing)
- summer holidays in warmer countries (accommodations in holiday cottages, less activity-oriented)
- pilgrimages (traditional tourism, particularly in the Himalayans)

(cf. SINGH 1992; ESCAP 1992; etc.)

Nature tourism primarily involves the following types of ecosystems, along with unspoiled coastal and mountain areas:

- tropical rain forests (animal and plant observation, nature photography);
- steppes and deserts with large-mammal populations (animal watching and photography);
- freshwater ecosystems and wetlands (boating, bird-watching, nature photography);
- nordic/(ant)arctic ecosystems (adventure activities, trekking).

Major nature-tourism regions are, apart from the mountain regions mentioned above, Northern Europe, North America, the rain forests of Latin America (Costa Rica, Ecuador, Brazil) and South Asia (Indonesia, Malaysia, Thailand), Australia and New Zealand as well as eastern and southern Africa (Kenya, Tanzania, Zimbabwe, Republic of South Africa). Moreover, an increasing role is played by the Antarctic and the mountains of Central Asia (cf. AGÖT 1995). In some of these countries nature tourism is already a significant market segment.

At the global level, nature tourism has thus far been more of a marginal phenomenon, but more and more frequently the countries where conventional beach tourism is prevalent can be observed to be marketing nature and particularly nature reserves as an *additional* attraction to their otherwise interchangeable touristic product. A by now classical example is the combination of beach and safari tourism in Kenya. As of recently, this is the strategy followed by the Caribbean Tourism Organization (cf. CTO 1991–1993), but also, for example, by Mauritius (RAMSAMY 1992) and Fiji (KING/WEAVER 1993). In Europe, Spain is going to particular lengths to add more variety to its touristic product by developing tourism in locales of natural beauty and rural areas.

2.2.3 Trends

In regard to qualitative market development, the WTO predicts the following areas of growth (cf. *WTO News* 4/1994 and 5/1994):

1. In future there will be more demand for pleasantly laid-out *holiday facilities* offering a number of programmes and activities, while at the same time shielding the visitor from any negative social and ecological conditions in the host country, than for urbanised tourist centres. From the environmental standpoint this trend should be viewed positively in part, if the holiday facilities are located in the "right" places, since such holiday facilities are usually "centrally" planned, i.e., by one touristic sponsor (organising company) and in this way protective steps can be more easily realised.
2. Many tourists will more frequently take several *short holidays* instead of a few long trips. This will entail an increased need of and higher demand on transportation infrastructure and increased global energy consumption for transportation. This will further bolster the trend to air travel, which is already evident in pressing demands coming from the tourism industry for expansion of existing airports or construction of new ones.

3. Demand will become increasingly critical and specialised. Coastal tourism will indeed remain the main segment of global tourism, but *nature* and *cultural tourism* will take on greater significance (see above). With it comes a higher environmental awareness on the part of the sources of demand, but this will also result potentially in a touristic stress on relatively untouched nature areas.
4. With growing quality expectations and volatility of demand and the increased price competition in the tourism industry, there is growing pressure on the host countries either to operate more cost-efficiently or to offer high-quality "holiday products". For the environment this can be an opportunity, as high environmental quality will become an increasingly important criterion in the choice of destination. However, in destinations which are already under heavy stress, there is the danger that expensive long-term *investments in the environment* won't be worth it if for the same or a lower price completely new "untouched" destinations can be developed.
5. *Incentive* and *conference tourism* will increase and at the same time raise the level of demand on a destination's "experience potential". This could result, at least in part, in a shift of business travel away from the cities to easily accessible, attractive coastal and mountain regions.
6. Another growing branch of tourism is *cruises*, so that from this quarter, too, increased stress on marine and coastal ecosystems is to be expected.

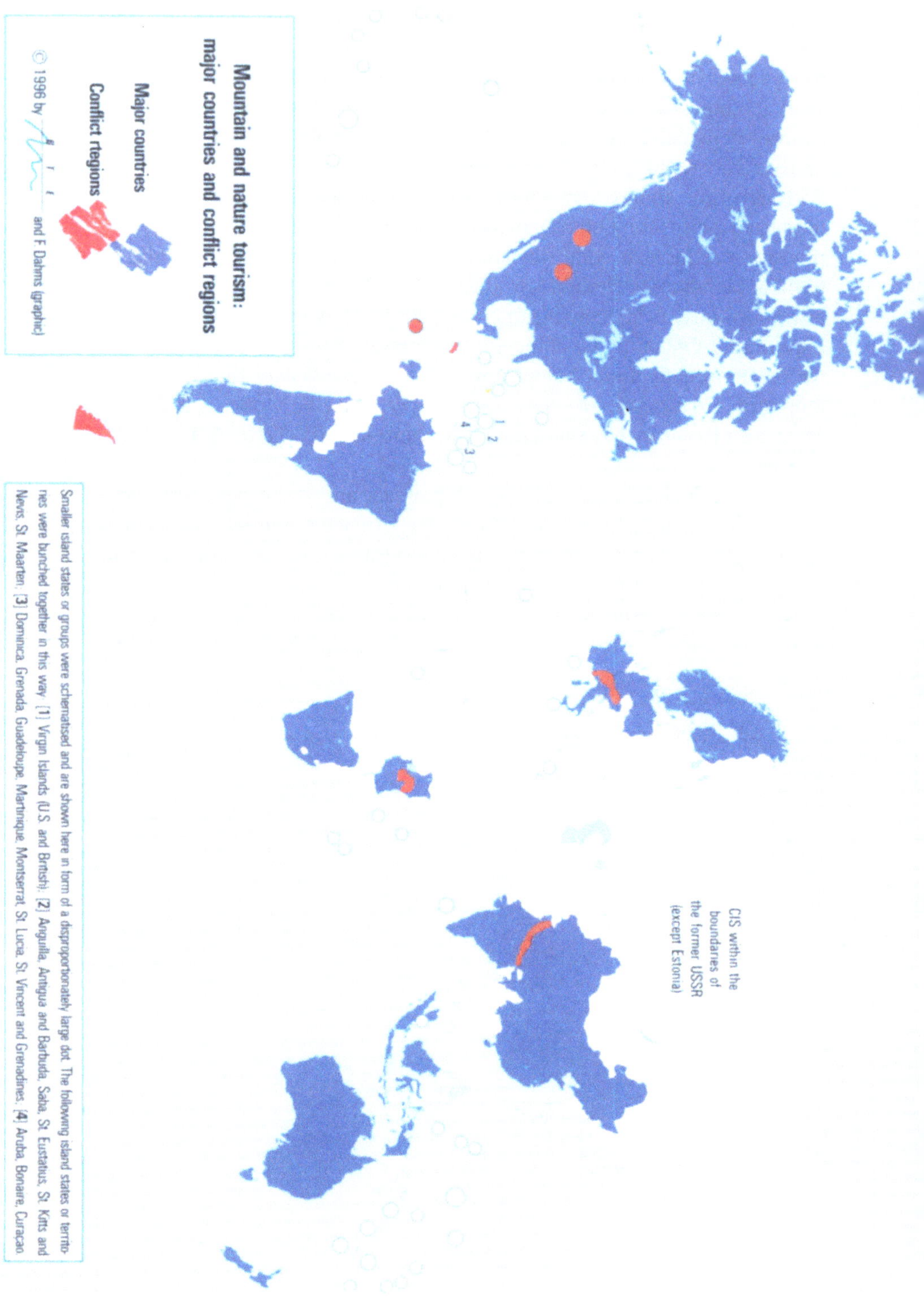

Map 4. Mountain and nature tourism: major countries and conflict regions.

3 Impacts of Tourism on Species and Ecosystems

It is undisputed and has been proven in numerous studies that in places where tourism takes place in massive form or encroaches on particularly sensitive ecosystems, severe impacts on nature and the environment result. This proof will therefore not be repeated here. The focus will instead be on systematising the well known impacts, demonstrating which are the major impacting factors, and identifying the impacted ecosystem types.

In regard to the impacting factors, tourism can be subdivided into the following *subsystems:*

1. recreational activities (main and complementary acitivities);
2. recreation infrastructure (paths/trails, sports facilities, pleasure-boat marinas, cable cars etc.);
3. infrastructure for accommodations, restaurants/cafés/bars and additional services (e.g., shops);
4. basic infrastructure (development and securing of infrastructure, transportation, utilities, waste disposal);
5. indirectly induced developments (regional migration, urbanisation, changing values etc.)

(adapted from ESCAP 1992, p. 7).

In addition, the differing spatial and temporal *levels of these effects* should be borne in mind. The spatial distinction should be made here between environmental impacts on the touristic site itself and those which turn up in other places (e.g., unregulated garbage disposal on the open sea or in the island's interior) and also between short-, mid- and long-term effects (e.g., blasting of coral reefs or their gradual destruction from eutrophication) (cf., e.g., HERRMANN et al. 1990).

The main focus will be on coastal tourism. In mountain tourism only the impact of winter sports will be dealt with. The impact of nature tourism is taken from a study on “Ecotourism as an Instrument of Nature Conservation?” (AGÖT 1995).

Map 5. Tourism and biodiversity in coastal regions: current potential conflict areas.

3.1 Impacts of Coastal Tourism

Impacting activity	Impacting factors	Impacted ecosystems
Sunbathing, etc., picknicking	Litter, fecal matter	**Sandy beaches, dunes:** changes in plant community through eutrophication, fire hazard, threat to animals
	Trampling on and breaking off of plants	Soil erosion, damage to vegetation
	Physical presence, noise	Scaring off of shy animal species (e.g., sea turtles laying eggs)
Swimming	Water contamination from sun-tan oil, soap	**Coastal waters, lagoons:** eutrophication
Non-motorised water sports (surfing, sailing, paddling)	Physical presence, movement	Coastal waters, wadden sea, beaches: scaring off of shy animal species (seals, water birds)
Skin diving *(see also boating)*	Damage to corals	**Coral reefs:** damage to reefs, shifts in species makeup
	Underwater hunting	Decimation of fish species, shifts in species makeup
	Stirring up of sediment	Decreased photosynthesis due to clouding of water
	Touching and feeding of fish	Shifts in species makeup, scaring off shy fish species
	Littering	Eutrophication, threat to animals (e.g., turtles, dolphins)
Motorised water sports (motor boats, water skiing, jetskis, parasailing)	Noise	**Coastal waters, lagoons, river mouths, wadden sea:** scaring off of shy animal species (e.g., water birds, seals)
	Wake waves, vibrations, stirring up of sediment	Damage to shore and underwater vegetation
	Mechanical effects of propellors	Injuring/killing animals (e.g., turtles)
	Contamination by oil and petrol, anti-rot coating	Water contamination (e.g., by heavy metals), poisoning of animals and plants
	Anchoring	**Coral reefs, eelgrass meadows:** mechanical damage
Sightseeing (with underwater or glass-bottomed boats)	Wake waves, stirring up of sediment, propellor effects, chemical contamination	**Coral reefs:** *see motor boats*
Fishing, clam diving	Overfishing, overgathering of particularly attractive species	**Open sea, coastal waters, lagoons, river mouths, beaches:** decimation of species
Nature observation (on foot or in boats)	Physical presence, noise	**Sand banks, rock cliffs, wetlands, mangroves:** scaring off of shy animal species (birds, reptiles, seals)
Walking, bicycling	*See sunbathing, picknicking*	Dunes, rocky cliffs, hinterland *see sunbathing, picknicking*

Table 8.1. Impact of recreational and other activities. (Source: *see Table 8.2)*

Impacting activity	Impacting factors	Impacted ecosystems
Sports (e.g., motor boating, horseback riding, golf)	Noise, wake waves etc., *see also motorised sports*	Mechanical damage, *see also motorised sports*
Eating, drinking	Overfishing of particularly tasty fishes and seafood	**Open sea, coastal waters, seafloor**: decimation of fish species, lobsters, clams
Purchase of souvenirs	Corals, clams	**Coral reefs, seafloor**: decimation of coral and clam species
Cruises	Illegal dumping of waste, sewage, oil and petrol	**Open sea**: endangerment and poisoning of animals and plankton
	Anchoring (particularly by small yachts)	**Coral reefs, eelgrass meadows**: mechanical damage
Visits to natural reserves	*See nature tourism*	*See nature tourism*

Table 8.2. Impact of recreational and other activities. Source: calculations based in part on ECTWT 1990, EUCC 1994, ESCAP 1992, STACHOWITSCH 1992, CHAPE 1990, SCHEMEL/ERBGUTH 1992, VISSER/NJUGUNA 1992, ISHMAEL 1993.

Impacting infrastructure	Impacting factors	Impacted ecosystems
Paths, beach promenades, boardwalks	Area covered over, treading, litter alongside, construction such as concrete poured on rocks and onshore sand-pumping	**Beaches, dunes, wetlands**: disruption of habitats, shifts in species makeup
Small-boat marinas and harbours for large passenger ships (cruises)	Bulldozing of shallow coastal segments	**Lagoons, estuaries, mangroves, salt meadows**: destruction of seafloor flora and fauna, shifts in species makeup
	Blasting of boat passageways	**Coral reefs**: destruction of habitats
	Harbour expansion for cruise ships	**Small oceanic islands (and above-mentioned habitats)**: destruction of habitats
Buildings	Overbuilding, sealing off of ground	**Beaches, dunes, rocky coasts**: destruction of habitats, disruption of land sea connections (e.g., sea turtle nesting spots)
	Clearing projects	**Dunes, hinterland, coastal waters, coral reefs**: soil erosion and sedimentation in the sea, destruction of plant communities
	Drainage and landfill in wetlands (also for reducing mosquito population)	**Wetlands, mangroves**: habitat destruction or severe impairment
	Extraction of building materials (sand, limestone, wood), extraction activity etc.	**Sand and pebble beaches, coral reefs, mangroves, forests in hinterland**: destruction of habitats, increasing erosion, deforestation

Table 9.1. Impact of recreational, accommodation and basic infrastructure. (Source: *see Table 9.2)*

Impacting infrastructure	Impacting factors	Impacted ecosystems
Parks, sporting facilities	Introduction of alien species	**Dunes, hinterland**: displacement of endemic species
	High water consumption for lawns, golf courses etc.	**Small islands, wetlands, arid hinterland**: increasing aridity, salinisation
	Use of fertilisers and pesticides	**Wetlands, dunes, coastal waters**: eutrophication, water contamination
Energy supply	Electric conduits	**Dunes, hinterland**: threat to birds
	Diesel generators: noise, exhaust fumes, oil pollution	**Beaches, dunes, hinterland**: disruptive effects on animals, water and soil contamination
Water supply	High water consumption by tourists and for parks etc.	**Small oceanic islands, freshwater wetlands**: habitat destruction by aridity or influx of salt water
Garbage disposal	Unregulated garbage removal	**Open sea, seafloor, dunes, wetlands, hinterland**: eutrophication, danger to animals
Sewage disposal	Inadequate sewage-treatment facilities	**Seafloor, coastal waters, coral reefs, eelgrass meadows, beaches, open sea**: clouding of water, algae bloom, oxygen deficit, death of large numbers of organisms
Transportation infrastructure	Building of airports (sealing off of land, landfills)	**Small oceanic islands, hinterland, rock coasts, wetlands, mangroves**: destruction and cutting off of habitat
	Operation of airports (noise, exhaust fumes, kerosene)	Impairment of habitats by soil and water contamination, disturbing of animals (especially birds)
	Road building (sealing off of land, landfill)	**Hinterland, rock coasts, dunes, wetlands**: cutting off of habitats
	Motor-vehicle traffic (noise, exhaust fumes, oil, petrol)	Habitat impairment by soil and water contamination, disturbance and running over of animals
Coastal-protection projects (breakwaters, moles); **pumping sand on shore** (usually as a result of previous intrusions)	Changes in currents	**Sandy beaches**: changes in habitats

Table 9.2. Impact of recreational, accommodation and basic infrastructure. Source: calculations based in part on ECTWT 1990, EUCC 1994, ESCAP 1992, STACHOWITSCH 1992, THOMAS 1990, CHAPE 1990, SCHEMEL/ERBGUTH 1992, VISSER/NJUGUNA 1993, ISHMAEL 1993.

3.2 Impacts of Mountain Tourism

Impacting activity	Impacting factors	Impacted ecosystems
Alpine skiing	Furnishing of technical infrastructure	**Slopes**: impairment of habitat by sealing off of ground, erosion
	Construction of ascent trails and ski runs	**Mountain forests and meadows, knee timber zone**: total change of habitats due to clearing, bulldozing
	Ski-run maintenance	Massive shifts in species makeup by eutrophication, mechanical damage, erosion
	Ski operations	Mechanical damage, erosion, expulsion of shy animals
Cross-country skiing, ski hiking, variant skiing	Skiing away from runs and cross country courses	**Mountain forests and meadows, knee-timber zones, marshes**: expulsion of shy animal species (e.g., grouse)
Aerial sports (hang gliding, paragliding)	Flying motions	**Inadequate mountain areas, rocks, open spaces**: expulsion of shy animal species (birds, mammals)
Free-time accommodations, huts and lodges	Situation at preferred sites	**Bodies of water, sunny slopes**: impairment of vegetation at water's edge and of thermophilic plant and animal species
	Inordinate transportation infrastructure due to remoteness of building	**Valleys, slopes**: impairment of ecosystems by sealing off of ground, cut off effects, erosion
	Wood consumption (for building and fuel)	**Mountain forests, knee-timber zones**: impairment, destruction
	Inadequate sewage treatment	**Oligotrophic mountain streams and lakes**: shifts in species makeup by eutrophication
	Unregulated garbage disposal	**Bodies of water, gorges, plant communities poor in nutrients**: shifts in species makeup by eutrophication, danger to animals

Table 10. Impact of winter sports, aerial sports and free-time accommodations on mountain ecosystems. Source: SCHEMEL/ERBGUTH 1992, SCHEMEL/RUHL 1980, SINGH 1992, altered.

3.3 Impacts of Nature Tourism

Tourist activities	Impacting factors	Impact
camping, picknicking	reclining, trampling	soil erosion and solidification, damage to vegetation from trampling
	noise	disturbing effect on animals (see above)
	garbage	eutrophication, threat to animals, impairment of landscapeis appearance
	gathering of firewood, campfires	biotope destruction, erosion and removal of nutrients (particularly above tree line), danger of forest fires
	washing in bodies of water with soap	water pollution and eutrophication
purchase/removal of souvenirs	purchase of feathers, animal parts, living animals	decimation of rare species
	purchase/removal of corals, seashells	destruction of reefs, species decimation
Adventure tourism		
mountan climbing/trekking	physical presence, trampling	damage to vegetation from trampling, disturbance to animals
	bored holes, chalk traces	damage and visual impairment of rock
skin diving	breaking off of corals	reef damage
	underwater hunting	decimation of fish species
water hiking (canoe, kayak)	physical presence	*see boating*
	landing	damage to shore vegetation from trampling
rafting	launching and landing of rafts	soil erosion and solidification, damage to shore vegetation from trampling
	transport of rafts	*see Table 12: cross country vehicles*
aerial sports	presence of aerial devices	disturbance of animals
	transport of aerial devices	*see Table 12: cross country vehicles*
Consumptive nature tourism		
hunting	exceeding kill quotas	decimation of animal species
	hunting outside allotted areas	decimation of animal species, massive disruptive effects
	violation of hunting ethics	cruelty to animals, rotting of meat
angling	overfishing	decimation of fish species, shift in species makeup

Table 11. Environmental impact of tourist activities in nature reserves. Source: AGÖT 1995, altered.

Tourist services/ infrastructure	Impacting factors	Impact
Activities of guides/drivers		
	illegal hunting	*see Table 11*
	fishing with dynamite	massive disturbance of aquatic ecosystems
	clearing of paths destruction of vegetation for demonstration purposes	destruction of vegetation
	chasing animals with cross country vehicles	massive disturbing effects, damage to vegetation, soil erosion
Transportation		
construction of transportation infrastructure (roads, airstrips, boat landings)	site coverage clearing of forests	deforestation, vegetation damage cutting off of integral ecosystems (e.g., impairment of animal wandering) sealing off of plots of ground
buses and cross country vehicles	driving cross country	soil erosion and solidification, damage of plants, running over of animals
	noise	disturbing of animals
	exhaust fumes, spilling of petrol and oil	air pollution, soil and water contamination
airplanes	noise, presence of airplanes	fleeing of animals
	exhaust fumes, spilling of oil and kerosene	air pollution, soil and water contamination
motorboats	noise	disturbance of animals (particularly water birds)
	exhaust fumes, spilling of petrol and oil	air pollution, contamination of water
	wake waves underwater vibration sediment turbulence	damage to shore vegetation and birds' nests impairment of water vegetation and fauna
Accommodations facilities		
construction of buildings	clearing, cutting of lumber	deforestation
	noise	disturbance of animals
	drainage	sinking of groundwater level, impairment of moist biotopes (e.g., mangrove forests)
	exposed sites unadapted architecture	disturbance of landscape appearance
accommodations operations	presence of people	disturbance of animals
	energy supply, diesel generator driven water pumps (noise, exhaust fumes)	disturbance of animals, water and soil contamination, air pollution
	water consumption	in dry areas or seasons sinking of ground water level, drying up of bodies of water on surface
	unregulated garbage disposal	*see Table 11*
	untreated sewage	contamination, eutrophication of ground and surface water

Table 12. Environmental impact of tourist services and infrastructure in nature reserves. Source: AGÖT 1995, altered.

3.4 Identification of Major Impacting Factors

Tables 8 to 12 summarise the most important environmental impacts from coastal, mountain and nature tourism, without for the moment assessing the *degree* of impairment. This depends on the intensity of the intrusion as well as the sensitivity of the impacted ecosystem and must therefore be ascertained on the basis of the individual case – e.g., with an environmental-impact assessment (EIA) – if a sound assessment of the expected impact is to be arrived at. Nevertheless, a few general statements can be made regarding the major impacting factors. Thus, most of the studies clearly show that the more severe impacts of tourism on species and ecosystems stem from the infrastructure and the building activity it entails rather than from the recreational activities themselves. This is particularly true of coastal tourism at massive concentrations. In contrast, with nature tourism, which needs relatively little infrastructure, the activities themselves are more in the foreground. Moreover, many conflicts are unavoidable by virtue of the choice of site and can often only be defused during the operational phase to a limited degree. The following statements thus refer to the touristic subsystems:

1. *site choice and development*
2. *operation of touristic infrastructure*
3. *touristic activities*
4. *indirectly induced effects.*

Re 1: For touristic facilities *attractive landscape sites are preferred:* on coasts primarily sandy beaches and dunes, lakes and rivers in the interior and, in the mountains, exposed mountain tops and slopes. As these are often transitional zones, valuable species-rich ecosystems are generally found there which are either destroyed outright or severely impaired by the buildings erected over them. In many countries hotels were (and still are) built right at the high-water mark of sandy beaches and/or in the dune belt behind them. In the western Mediterranean a large portion of the coastal dunes were destroyed in this way (cf. UNEP 1992; EEA 1995; etc.).

Secondly, *non-adaptation to existing natural site conditions* should be mentioned. This is particularly true of coastal wetlands, which, for lack of more suitable sites, are drained and filled in for the construction of buildings, roads or airports. Particularly frequently affected are small islands, especially in the Caribbean. A second important impacting factor in this regard is marinas, which are preferably built on shallow coasts, e.g., in lagoons or at river mouths, because there they are better protected from the open sea. On the other hand the water is often not deep enough there, so that dredging is required, permanently damaging the impacted ecosystems. In some countries (e.g., Fiji, Dominican Republic) boat passageways were blasted in the coral reefs situated before the coast. Mangrove forests, a land-sea transitional zone, are particularly frequently impacted by both development types (cf. CHAPE 1990; INSKEEP/KALLENBERGER 1992; etc.).

Thirdly, *building materials* are often removed from ecosystems in a non-sustainable way. This is particularly true of the fine sand of beaches, which is used to mix concrete. This increases the danger of erosion on the beaches, so that in some cases sand is pumped onshore and coastal-protection steps have to be taken. But the use of *traditional* building materials such as wood or reef limestone can also pose problems when this use is excessive.

Re 2: The most severe impairments are caused by untreated sewage, inadequate garbage removal and excessive water consumption. Many authors (e.g., STACHOWITSCH 1992) are convinced that pollution from *sewage* is one of tourism's biggest problems, as it can scarcely be confined spatially, particularly in the sea, and the changes in the nutrient balance it causes inflict extensive damage on the impacted habitats. This is particularly true of oligotrophic mountain streams and the in this regard very sensitive coral reefs. In contrast, naturally nutrient-rich ecosystems such as, for example, mangroves can perform important buffer and filter functions to a limited extent.

Garbage is also a major problem. In developing countries, in particular, there are hardly any capacities for regulated disposal. In this regard, illegal dumping in the interior, as contrasted to dumping at sea or sinking in the deep sea, as practised in the Maldives, is less harmful, as eutrophication and contamination from harmful substances is limited to a relatively small space. The problem is particularly acute on small islands, where suitable dumping sites scarcely exist. In this context cruise tourism in the Caribbean should be mentioned, whose large passenger ships far exceed the capacity of the islands and not infrequently dump their waste on the open sea (ISHMAEL 1993).

Water consumption by tourists and tourism facilities in Kenya, for example, at 300 to 500 liters per person per day, amounts to ten times the minimum requirement arrived at by UNICEF (VISSER/NJUGUNA 1992). The least portion of this amount is taken up by drinking water. Water is used primarily for showers, swimming pools and the watering of gardens and golf courses not typical to the locale. The problem primarily occurs in arid climates and on small islands with limited water supply, but also at many destinations with more plentiful precipitation which are frequented by tourists preferably in the dry season. This results not only in social conflicts (e.g., with local farmers in Tunesia or Goa), but also in the fact that wetlands dry out and salt water intrudes into near-coastal freshwater biotopes (e.g., in Florida and on some islands of the South Pacific; cf. DUGAN 1993).

Re 3: Whereas many traditional touristic activities are concentrated on sandy beaches and the touristic agglomerations, more recent leisure-time trends tend towards athleticism and *increased mobility*. This means that often relatively unspoiled areas, particularly in the vicinity of the agglomerations, but also at more distant locations, are made accessible which up to now were largely untouched by tourism. These types of tourism – called sports, adventure and nature tourism – generally take place in less massive form, but at certain points they can lead to considerable environmental stress.

Various kinds of *aquatic-sports activities* should primarily be named here, such as sailing, surfing, waterskiing or motorboating. These activities are already being indulged in on a massive scale and will continue to be very popular. In particular, motorised activities constitute an increasing stress on freshwater ecosystems and near-coastal biotopes. *Skin diving* has especially high growth rates, and it is particularly significant for global biodiversity, as it is preferably practised at coral reefs. Some islands, such as the Maldives or Saba, in the Netherlands Antilles, direct their advertising almost exclusively at this target group.

Another trend sport worldwide is *golf*. In Southeast Asia, in particular, a veritable boom of this activity is taking place. Golf courses need large quantities of water, fertilisers and pesticides to protect the generally non-indigenous grass from prevailing environmental influences.

In mountainous regions *alpine skiing* constitutes one of the greatest touristic stresses, due to its high demands on the infrastructure. Due to climatic changes which appear to be occurring in the Alps, the most important winter-sport region in the world, there is growing pressure to develop other mountainous areas, particularly in the Northern Hemisphere. In addition, *trekking* and *mountain climbing* are growing demand segments which will increasingly constitute a conflict potential at certain points, in particularly attractive landscapes high in the mountains (cf. STRASDAS 1994).

Nature tourism can also take place on a massive scale at certain points, with the corresponding impairments, especially wherever protected areas are near tourist centres on the coast. East African national parks are particularly affected, whose large-mammal populations constitute a particular attraction for *photo safaris* (cf. AGÖT 1995).

Re 4: Large-scale touristic projects can have considerable distorting effects on the national economy, especially in the developing countries, when economic subsystems more typical of highly advanced service economies are introduced to agrarian economies abruptly and in massive form. This has not only social consequences, but ecological ones, as well. Because of the huge income gradient, *migration* from rural areas to the tourist centres, which are typically on coasts, is a frequent consequence. Two examples of this are Turkey and Kenya (VISSER/NJUGUNA 1992). Similar effects are known to have been caused even by nature tourism (e.g., Galápagos Islands in Ecuador; national parks in northern Tanzania).

The increased population density leads in the affected areas to further *environmental strain* (particularly resource depletion, sewage and garbage), which in part exceeds that of tourism, as it is more difficult to control. Thus, in the vicinity of holiday centres (some of which are uniformly planned and equipped with sewage-treatment facilities of their own), a number of small enterprises of the informal sector (restaurants, souvenir shops etc.) are often found which pose much greater problems for the environment, as they usually have no technical equipment for sewage treatment and garbage disposal. Examples for this may be found in many developing countries, e.g., in the Dominican Republic (INSKEEP/KALLENBERGER 1992) or in Malaysia.

On the other hand, such economic shifts can also result in a *decrease of environmental stress*. Examples are island states far away from markets, such as Fiji or Mauritius, where tourism is increasingly replacing environmentally harmful, hardly profitable agricultural and forestry enterprises (export of sugar cane and timber; CHAPE 1990). A similar potential is also inherent to nature tourism, particularly in rain-forest regions and areas with large-mammal populations, where their touristic marketing can at least partially supplant the felling of timber, placing land under cultivation or poaching (cf. AGÖT 1995).

The environmental effects of tourism which go beyond the local and regional level should also be mentioned, in particular those caused by the *airplane*, the major means of transportation. As of today, an estimated two thirds of global air traffic is taken up by tourism. It is true that this has hardly any direct adverse effect on biodiversity (except at the airports themselves; see above), but indirect effects are to be feared from the influence on the global climate (cf., e.g., UNEP 1992).

3.5 Identification of Major Ecosystem and Tourism Areas

In considering the statements made in the previous chapter from the standpoint of the preservation of the diversity of species and habitats, the following valuable *ecosystem types* may be named which are particularly severely impacted by tourism:

A. Coastal tourism

1. *Oceanic islands* (diverse ecosystems in a confined space, high degree of endemicity, shortage of resources, high sensitivity, very intensive tourism);
2. *coral reefs* (very high species diversity, high sensitivity, high tourist attractiveness);
3. *offshore waters* (in the tropics high biodiversity, intensive touristic zone, particularly strongly affected by pollution);
4. *mangroves* (high productivity and coastal-protection function, particularly strongly impacted by touristic construction and development measures);
5. *near-coastal wetlands* (primarily lagoons; high proportion of rare species, very frequently affected by development and drainage projects);
6. *sandy beaches* (the most intensively frequented touristic zone; relatively few species, but at certain points part of the habitat of endangered species, e.g., sea turtles);
7. *coastal dunes* (the second most highly frequented touristic zone – after beaches; relatively few species, many of which are rare, however; high sensitivity).

B. Mountain tourism

8. *High mountains* (increasing touristic frequency; relatively few species, many of which are rare and some endemic; very high sensitivity).

C. Nature tourism

9. *Freshwater ecosystems* (including adjoining wetlands, high touristic frequency, high proportion of rare species, sensitive to pollution);
10. *spectacular protected areas* (primarily World Heritage Sites, parks with large-mammal populations; increasing tourism, refuges for endangered animal species);
11. *(ant)arctic coastal ecosystems* (increasing tourism at certain points; relatively few species, many of which are rare; extremely high sensitivity).

Map 1 shows those *countries* of high significance for global biodiversity. It is based on the following criteria (adapted from IUCN):

- particularly high species diversity
- significant presence of tropical rain forests
- islands with numerous endemic species
- countries with major coral reefs.

As to global regions, conflict potential between tourism and biodiversity is mainly concentrated in the following regions:

- *Southeast Asia* (highest significance for global biodiversity, world's highest tourism growth rate; impairments are even now emerging);
- *Caribbean* (world's most intensive touristic frequency per unit of area; hardly any economic alternatives; coral reefs, high significance for endemic species; marked impairments already exist);
- *Central America* (vey high significance for global biodiversity, hardly any impairments, but very high tourism growth rates);
- island groups in the *Indian Ocean* (generally with intensive tourism per unit of area; hardly any economic alternatives; coral reefs, endemic species; impairments are beginning to appear);
- island groups in the *South Pacific* (frequently with intensive tourism per unit of area; hardly any economic alternatives; coral reefs, endemic species; impairments are beginning to appear);
- *Europe:* Mediterranean and Alps (world's highest touristic frequency and impairment, but on a global scale relatively low species diversity).

In addition, there are conflict areas in certain areas of *East Africa* (Kenya, Tanzania), in the *Himalayans* (Nepal, India) due to intensive nature tourism, and in *South Africa*, which has very high tourism growth rates. *Australia* is one of the most important countries for global biodiversity (endemic species, world's largest coral reef) and is very highly frequented by tourism in places. However, conflicts are apparently minimised by increasingly significant nature-conservation policies and the development of sustainable tourism. In the likewise important countries with tropical rain forests in *South America* and *Africa* there is comparatively little tourism. In this case the dangers to the ecosystems emanate for the most part from non-touristic factors. In these countries tourism can be viewed as having conservation rather than conflict potential. The case is similar in regard to the highly endangered endemic fauna of *Madagascar*.

4 Solution Strategies

Criticism of the social, cultural and ecological consequences of tourism was first voiced back in the 70's. But more than a decade elapsed before this insight became widespread and was generally accepted (cf. BMZ 1993). Among the people involved with tourism today it is almost a commonplace that tourism can have adverse environmental impacts and that they should be limited as far as possible. Solution strategies in the form of programmes, planning strategies, regulation and recommendations can thus be found in governmental and inter-governmental bodies and non-governmental organisations as well as in the tourism industry itself. These plans are of varying – generally slight – degree of feasibility or binding quality, which is evidenced, for one thing, by the fact that there is a multitude of recommendations, but there are hardly any binding agreements. Most radical in this regard are legislative regulations in a number of destination countries, but very frequently – particularly in developing countries – there is no way to implement or enforce them effectively.

4.1 Solution Strategies of Governmental and Intergovernmental Bodies at the Global Level

No formal international agreements bringing nature conservation and tourism into a common context exist at present. However, the Convention on Biodiversity and Agenda 21 are approaches which have moved a number of intergovernmental organisations and international development co-operation to come up with initiatives geared to arriving at sustainable forms of tourism.

The most imporant intergovernmental organisation in the tourism industry is the *World Tourism Organization* (WTO). The WTO was founded in 1975 with the principal aim of promoting international tourism by establishing suitable framework conditions and removing existing barriers.

The economic effects, particularly in the developing countries, were and continue to be its principal concern. However, for some time now the environmental compatibility of tourism has been playing an increasingly prominent role at the WTO, as well. This task is the responsibility of the Environment Committee, one of five committees subordinate to the Executive Council (PAHR 1987).

The WTO co-operates closely with the United Nations, especially with UNDP (United Nations Development Programme) and UNEP (United Nations Environment Programme). Among the results of this co-operation have been regulations for tourism in protected areas (WTO/UNEP 1992), the Ten Commandments for Protecting World Heritage Sites (DROSTE et al. 1992), "Principles of Sustainable Tourism" (WTO 1993, p. 40) and recommendations for planning at national regional and local levels (op. cit. and WTO 1994b). In co-operation with the World Travel and Tourism Council (WTTC), the WTO has developed a strategy for applying the objectives set forth in Agenda 21 to the travel and tourism industry: "Towards Environmentally Sustainable Development".

In international *development co-operation* (DC) tourism has hardly figured in its conventional form (i.e., non-adapted large-scale projects) since the 80's, as its social, economic and ecological effects are thought to be predominantly detrimental. In contrast, the *nature-tourism* segment is being given increasing attention in connection with nature-conservation projects. The most advanced in this regard are the activities of the USAID (United States Agency for International Development), which is conducting a number of pilot projects, particularly in Latin America and Africa. However, a comprehensive plan for sustainability of tourism has yet to be worked out. Initial steps in this direction are contained in the Global Environment Facility programme of the World Bank and in German and Scandinavian DC (AGÖT 1995, p. 89f.). An important role is played in this regard by international non-governmental organisations, but thus far only the WWF (World Wide Fund for Nature) has come up with a coherent plan or guidelines (see chapter 4.4).

4.2 Solution Strategies of Intergovernmental Organisations at the Regional Level

Regional nature-protection agreements in which tourism plays a part (Alps Convention, Antarctica Treaty, Convention on the Protection and Development of the Wider Caribbean Region, South Pacific Regional

Environment Programme and various European agreements) are analysed and assessed in the section of this study dealing with the legal aspects (Section D, chapters 11ff.). Solution strategies in Europe which deal with tourism and its impact on nature and the environment are also examined against the backdrop of the nature-conservation expertise of the present study (chapters 5ff., especially chapter 10). It should be borne in mind that many of these agreements were reached within the framework of the UNEP's *Regional Seas Programme*, and thus agreements on coastal tourism would do well to take them into account. A strong potential candidate for sponsoring regional programmes would continue to be the *Association of Small Island States*, which was very active at the Rio Conference (KNECHT/ CICIN-SAIN 1993).

Of interest at this juncture are the activities to date of intergovernmental organisations relevant to tourism in the areas of environmental protection and nature conservation. The most important organisations of this kind are the *Pacific Asia Travel Association* (PATA), comprising the states of East and South Asia, Australia, Oceania and North America, and the *Caribbean Tourism Organization* (CTO), in which 30 Caribbean states are represented as well as a number of tourism enterprises. Smaller touristic associations are the *Tourism Council of the South Pacific* and the *Mundo Maya* project of a few Central American countries. The main objective of these organisations, in which economic interests predominate, is regional marketing, but environmental criteria are playing an increasingly important role.

Thus, the PATA has published a "Code for Environmentally Responsible Tourism" and periodically holds conferences on ecotourism (UNEP 1992; CAMPBELL 1994). In its "Guidelines for the Integration of Tourism Development and Environmental Protection in the South Pacific", the Tourism Council of the South Pacific recommends that its member states bear natural capacity limits in mind, introduce EIAs for tourism projects, establish a system of protected areas etc. (CHAPE 1990).

The topic of ecotourism has been particularly frequently discussed in regard to and in the Caribbean: since 1991, the CTO has been regularly holding ecotourism conferences. The member states were urged to adopt a comprehensive tourism-environment agreement with which the recommendations of the first conference could become reality. Among its goals are the realisation of unified coastal-zone management and more stringent control of garbage and sewage disposal (CTO 1992; DOUGLAS 1992). However, implementation of the many recommendations hardly takes place in the individual countries. A particularly striking example is a resolution of the Organization of East Caribbean States to drastically increase the harbour dues for cruise ships, which are a source of particular environmental stress. With only one (!) exception, none of these countries adhered to the (non-binding) resolution. This island – St. Lucia – was consequentially no longer called at by the ships and incurred considerable economic losses (ISHMAEL 1993).

The example of the Caribbean is probably typical in a number of ways. This is a region of high touristic intensity which has to deal with increasing environmental impairment, which is costly, and has in part already resulted in a stagnation of demand or slight economic setbacks. Thus, environmental-protection efforts are on the one hand indeed in the vital interest of the region; on the other hand, ecotourism is seen rather as an *additional* niche to be developed which will add variety to the total product and bring additional growth, even if everyone is talking about ecological-capacity limits. Now it must be feared that hitherto inaccessible areas, e.g., in the interior of the island, will also be impacted by tourism.

It remains to be seen if the Mundo Maya project, with which Central American states are attempting to attract the tourism market to their historical Maya culture, will be more sustainable in this regard. Close co-operation with Paseo Pantera is planned, another supranational "nature-protection project". Here, too, the idea is not to "ecologise" existing tourism, but to develop new demand in the area of cultural and nature tourism (AGÖT 1995).

4.3 National Programmes for Sustainable Tourism

Binding government regulations to minimise the environmental impact of tourism exist most frequently at the national level. However, here the problem often is that these regulations are not implemented or cannot be monitored. In contrast, programmes of smaller island states such as, for example, the Cayman Islands, have a better chance of being implemented if a consensus is reached among all participants beforehand. Below a few exemplary countries will be described where relatively extensive programmes or solution strategies have been developed and in part implemented dealing with the general problem of tourism vs. environment.

The government of *Australia*, one of the most important countries for global biodiversity, established a number of working groups in 1990 to study the possibilities of sustainable development in various areas, one of which was tourism. The result was a very comprehensive national tourism strategy which was officially announced two years later by the Commonwealth Department of Tourism. This programme was recently presented in more detail in the area of nature tourism, including concrete implementation strategies. Thus, although only a portion of Australian tourism is addressed, it is a very significant one, as the coasts are also regarded as nature areas, with large portions under protection (e.g., the Great Barrier Reef off the northeast coast). For the implementation of the National Ecotourism Strategy a total of $10 million are set aside for the next four years, to be used in the following areas:

- evaluation and permit processing for touristic enterprises
- market research
- reduction of energy consumption and garbage
- construction of site-adapted and -preserving infrastructure
- environmental education and information
- research on methods of environmental monitoring
- pilot projects of integrated regional planning
- ecotourism training of tourism personnel
- conferences and workshops

(RICHARDSON 1993; CDT 1994).

Examples of solution strategies for *island states* with relatively highly developed tourism are Fiji, the Cayman Islands and the Maldives.

Fiji, which already began to develop touristically in the 60's, adopted as early as 1973 its Tourism Development Programme, which was funded by the UNDP and set forth a number of environmentally relevant goals, such as the importance of an intact environment, especially of the coral reefs, for tourism, the development of protected areas or, regarding hotel construction, setting a minimum distance of 30 m from the high-water mark. However, the implementation of these recommendations was a long time in coming. In 1989 the government enacted a Tourism Masterplan which contained major elements of sustainable tourism (CHAPE 1990; KING/WEAVER 1993).

The *Cayman Islands* and the Maldives are ecologically extremely fragile, as they consist exclusively of flat coral islands which can be quickly destroyed by unchecked development. On the Cayman Islands, which were already developed touristically in the 50's, the Marine Conservation Law was decreed in 1978, prohibiting the introduction of untreated sewage into the central lagoon and severely limiting the catch of lobsters and clams. Nevertheless, touristic development continued unabated, to the particular detriment of the mangroves. As a result, pressure was brought to bear, primarily by enterprises promoting skin diving, and a Marine Parks Plan was drawn up which divided the islands into various protected zones and areas, thus protecting the coral reefs from nearly all sedimentation (caused by the felling of the mangrove forests; BUSH 1993).

The tourism boom on the *Maldives* is more recent, but in this case, too, a tourism plan was drawn up in a relatively early phase (1983), which took the limited resources of the islands, some of which are tiny, into account. The plan contains regulations on the islands' maximum capacity, a ban on the use of reef limestone for building purposes, the adaptation of the infrastructure to sites and sea-current conditions, ban on underwater hunting and the collecting of corals and crustaceans (INSKEEP 1992; WTO 1994b).

In the *mountain tourism* sector, Bhutan's tourism planning is among the most exemplary in the world. Bhutan limited the number of foreign tourists who may travel in the country initially to 3,000 (1990). By 1996 a gradual increase to 6,000 was planned. Moreover, the country was divided into zones in which tourism is permitted in various degrees of intensity or not at all. Although the reason for the very restrictive management of tourism is explained by the government's intention to minimise socio-cultural conflicts, environmental standpoints were also major criteria. Thus, for example, already extant infrastructure is put to use wherever possible. New hotels have to adapt to existing architecture and include adequate sewage-disposal systems (op. cit.).

4.4 Solution Strategies of Non-governmental Organisations

Non-governmental organisations (NGOs) have been and continue to be the driving force of criticism of the ecological and social impacts of tourism. In 1980, long before the WTO began to deal seriously with the problem, the first conference criticising tourism was held in Manila on the topic of long-distance tourism, from which the Ecumenical Coalition on Third World Tourism (ECTWT) later emerged. Subsequently other *tourism-specific NGOs* were formed, particularly in the major "source" countries, e.g., Tourismus mit Einsicht (Insightful Tourism) in Germany, Arbeitskreis für Tourismus (Tourism Working Group) in Switzerland and Tourism Concern in Great Britain, all of which contributed to bringing the topic of the environment to the attention of the tourism industry and the governmental sector (BMZ 1993, p. 51f.).

Against the backdrop of nature-conservation work in the developing countries, which is increasingly difficult to find funds for, some of the large *nature-conservation organisations* began to take an interest in tourism, albeit almost exclusively in the nature-tourism segment. The most active in this regard are WWF and IUCN, but also a few organisations in the United States such as Conservation International or the Nature Conservancy. An NGO which is exclusively concerned with ecotourism is the *Ecotourism*

Society (TES), also located in the U.S., which has published guidelines for tour operators, for example, and is currently testing a procedure in Ecuador for evaluating environmental and social compatibility of tour plans (TES 1993, 1994).

One of the most comprehensive catalogues of demands on the tourism industry and others involved in tourism – one which goes beyond the scope of nature tourism – has been compiled by the *WWF* in co-operation with Tourism Concern ("Beyond the Green Horizon – Principles for Sustainable Tourism"). The ten principles can be sketched as follows:

- sustainable use of resources
- reduction of excessive consumption and waste
- preservation of natural, social and cultural diversity
- comprehensive tourism planning, including EIAs
- support of the local economy
- involvement of local population
- all-embracing co-operation between involved parties
- training of tourism personnel
- responsible promotional activity
- support of research

(WWF 1992).

An interesting initiative concerning a limited, but rapidly growing segment of global tourism, i.e., skin diving, is the *Coral Reef Alliance* (CORAL), located in the United States. CORAL initiates financial and political steps to activate skin divers and operators of skin-diving tours to protect coral reefs. A nature-conservation fund is geared to lend financial support to protected areas. Moreover, the initiative hopes to form a pressure group which can exert an influence on the politicians of destination countries by pointing out that members will only travel to those reefs whose protection is guaranteed (COLWELL 1995).

4.5 Solution Strategies in the Tourism Industry

In view of the institutional weakness of government authorities, particularly in the developing countries, and the difficulties intergovernmental organisations have in taking regionally co-ordinated action (see chapters 4.2 and 4.3), the self-regulation of the tourism industry takes on great significance. The industry has an immediate interest in environmental-protection measures in the following cases:

a) if costs can be reduced through them (e.g., saving energy or water);
b) if environmental damage would lessen the attractiveness of a destination (e.g., ocean pollution, traffic noise, garbage problems);
c) if it results in an enhanced image which can be used as an advantage over competitors.

As the first two conditions are more and more frequently the case and the last virtually always is (nature and environmental protection are viewed positively in virtually all of the source countries and demand groups), there have been numerous examples of initiatives on the part of the tourism industry to reduce the environmental effects of their business activity. They range from non-binding declarations of intent to chlorine-free-bleached paper for travel catalogues, quality seals for holiday sites down to concrete measures, e.g., to save energy and water as well as reduce sewage and garbage.

Solutions have been suggested by individual companies as well as regional and national organisations. Examples are the American Society of Travel Agents' "Ten Commandments on Ecotourism" (WTO 1993, p. 142), the Tourism Industry Association of Canada's "Code of Ethics and Guidelines for Sustainable Tourism" (UNEP 1992) and German Travel Agents' Association (DRV) project "Massnahmen zur Förderung des Umweltschutzes in der Touristik" (Steps for Promoting Environmental Protection in Tourism; DRV 1995).

At the global level, the *Green Globe* programme was recently inititiated by the World Travel and Tourism Council (WTTC). The WTTC, currently the world's leading association in the tourism industry, was founded in 1990 with the following main objectives:

- promotion of economic development through tourism, particularly in the developing countries;
- expansion of touristic infrastructure, particularly transportation infrastructure;
- promotion of growth of tourism and removal of growth impediments of physical (!) or bureaucratic nature or due to taxes (!);
- promotion of sustainable development, particularly through self-regulation of the industry;
- training of qualified tourism personnel

(WTTC, Media Information).

Green Globe is a logo for tourism enterprises which can be used by all applicants who declare an intent to *improve* their business practises regarding the environment. Membership in the Green Globe programme must renewed every year, requiring members to provide proof of their continued active commitment or concrete applications. The WTTC's Environmental Guidelines serve as a model, but a degree of objective fulfillment is not stipulated. The programme provides detailed information (worldwide data banks, brochures) and continuous counselling and continued-training services. Moreover, competitions are to be organised and prizes awarded for exemplary achievements in the environmental field (WTTC; Green Globe).

The Green Globe programme is highly controversial. On the one hand it is not content with simply providing a logo, but in addition offers a great amount of continued assistance. On the other hand it has several quite obvious weaknesses. Criticism by the Ecotourism Society, for example, is concentrated on the following points:

- the logo can be used as a quality seal, even though there are no minimum requirements and no surveillance of the operators'/companies' environmental commitment;
- there is no independent assessment of the applicants; the Board of Directors comprises – with the exception of the "eco-icon" Maurice Strong – only tourism executives;
- payments are forwarded directly to Green Globe, so that there is scarcely interest in turning down potent funders (i.e., large enterprises which must pay much higher dues)

(EPLER WOOD 1994).

Against the backdrop of the general limits to the environmental commitment of the tourism industry mentioned above and the WTTC's main objectives, a bit of skepticism about the industry's real willingness to regulate itself is definitely called for. It apparently interprets self-regulation primarily as the dismantling of government intervention and less as a (voluntary) acceptance of governmental control functions in the field of environmental protection and nature conservation. This is particularly true of restrictions on growth which can become imperative due to limited ecological capacity. The parallel objectives of nature protection and touristic-resources conservation so often cited by the industry will have reached its limits here, at the latest. The continuing commercial success of mass tourism, e.g., on the ecologically highly impacted Mediterranean coasts, shows that tourists generally don't stay away until they feel directly impacted. The loss of species diversity is usually not noticed by tourists and is thus only rarely detrimental to the tourism business.

Section B. The European Situation

Whereas in Section A the real and potential impacts of tourism on the biodiversity of individual bio-geographical areas of the world were examined on a global scale, the focus of the present section will be on the actual conflicts between tourism and biodiversity on European coasts. Europe needs to be examined more closely, for although in the rankings of the International Union for Conservation of Nature IUCN (1994b) the European region has a low biodiversity by global standards (cf. Section A), many of its coasts are of outstanding significance from a European standpoint in regard to biodiversity. At the same time, touristic pressure on Europe's coasts is disproportionately high, so that there is a conflict potential which would be ranked around mid-scale by global standards. With a share of 59.3% of international tourism, the European region accounts for about two thirds of global tourism (cf. WTO 1994a). One consequence of this is that the *biological diversity of European states is steadily declining* (European Environmental Agency – EEA – 1995). A drop in species diversity and a loss of intact, i.e., "functioning" habitats, from an ecological standpoint, may be observed everywhere in Europe.

"Many people are concerned about the Amazon or Antarctica. Closer to home we tend to neglect the threat to our own remaining natural landscapes. Along Europe's coasts a silent drama is being played out. Nowhere in the world has the coastal zone been so badly affected as in Europe" (European Union for Coastal Conservation – EUCC – 1992, p. 3).

One region of confrontation between biodiversity and tourism is the Mediterranean. In the Mediterranean region there are over 25,000 different plant species, over half of which are endemic (cf. GRENON/BATISSE 1991). However, the low ranking of its biodiversity by the IUCN should not lead to the conclusion that Europe's ecosystems are insignificant and need not be provided with special protection (cf. European Centre for Nature Conservation – ECNC – 1995). The IUCN rankings, as already mentioned, are purely quantitative statistical evaluations; the qualitative aspects of individual ecosystems are completely disregarded. The rankings can only function as an indicator of a need for action or the "urgency of action" on a global scale. As to the actual quality of the ecosystems, little can be said using a purely quantitative evaluation at the global level.

At the European level various initiatives are therefore demanding that the purely quantitative surveys by the IUCN be complemented by the special national idiosyncrasies of historically evolved, natural and culturally influenced landscape structures and nature areas (cf. ECNC 1995, Biodiversity Workshop Wageningen 4/95, Summary). *Every bio-geographical region in the world has its own typical makeup of nature areas which are characterised by their own specific balance and/or optimum. A naturally oligotrophic bog in Ireland, for example, cannot be ranked according to the number of species it contains, because, on the contrary, a great variety of species would indicate an "altered, disturbed" state in this habitat's highly specialised ecosystem, according to the EEA (1995).*

The pilot-study project now being conducted by the UNEP and EU (cf. Section A) evaluates information on land use, habitat structure and landscape types, as well as animal and plant species, with regard to biological and landscape diversity. Here the significance of a large, intact, boreal forest area or moor (both with relatively few species) should be just as great as the wealth of species of a "hot spot", a term denoting centres of limited size with an extremely large number of species such as those found, for example, in the Mediterranean region (WASCHER 1995).

The examination of the European situation in regard to tourism and biodiversity on the coasts will proceed according to the following steps:

1. description of the particular natural characteristics of European coastal and marine ecosystems (chapter 5);
2. identification of the most important regions for European species diversity and of significant ecosystem types in Europe (chapter 6) and description of the major causes of threat to the European seas (chapter 7);
3. description of the quantitative and qualitative characteristics of European tourism, identification of its impact and current major areas (chapters 7 and 8);
4. identification of the conflict potential currently existing in the European region between biodiversity and tourism (chapter 8);
5. description of illustrative examples, conflicts of interest and national solution strategies in selected European coastal regions (chapter 9);
6. description of solution strategies regarding development of tourism at the European level (chapter 10).

5 Coastal and Marine Ecosystems in Europe

In chapter 1.3 (Section A) a global overview of the most endangered sea-coast habitats was already provided. The European seas and coasts are characterised by a large number of ecosystem types which will be described below. Particularly sensitive and therefore endangered coastal areas in Europe are salt meadows and dune landscapes and the ecosystem of the Wadden Sea (mudflats) (cf. chapters 6, 7 and 9). An overview of the most important characteristics and peculiarities of European seas and coasts is in chapter 5.3.

5.1 Systems and Subsystems

The "nature system coast" can in general be subdivided into the subsystems coastal waters, near-coastal mainland, the shore fringe and fresh water (surface and ground water). The following brief *classification of coastal segments in Europe* follows the classification of the Ramsar Convention and the CORINE biotope project (cf. EEA 1995), coupled with an overview of the European Union for Coastal Conservation (EUCC) of 1992. Unlike most land ecosystems, the hydrological, chemical, biological and geological subsystems of coastal and marine ecosystems are subject to high *dynamics*. The European coastline has developed in the course of millions of years and steadily continues to change. Apart from isostatic changes still taking place, as a result of the last Ice Age, huge erosion and sedimentation processes are caused by various natural processes:

- by waves and tides,
- by geomorphological changes in the interior which are transported to river mouths by rivers,
- by meteorological factors such as wind and storms,
- by changes in sea level
- and by geological processes (e.g., seismic/volcanic activity).

The various types of coastal segments can be classified as follows:

5.1.1 Rocky Coasts

A large portion of the European coasts consists of rock: cliffs, fjords, rocky plateaus or islands rising steeply out of the sea. Rocky coasts are not rare in Europe, with a great variety of forms and types of rock. This is illustrated, for example, by the steep coast of Rügen, the chalk walls near Dover and Calais or the Blue Grotto on the Isle of Capri etc. – each coast is unique in appearance. In northwestern Europe, for example, the following types of rocky coasts may be distinguished (adapted from LIEDL et al. 1992):

skerry coasts: formed by free glacial erosion, i.e., by removal or abrasion of the rock by flowing ice – slightly wavy, humped relief;

fjord/skerry coasts: formed by free and directed glacial erosion – slightly wavy with drowned lower valley reaches;

fjord coasts: formed by directed glacial erosion – largely confined valleys with seawater;

cliff-row coasts: formed structurally, tectonically or by the sea – sometimes drowned platform steps with steep coasts.

5.1.2 Loose-Rock Coasts

Coasts of glacial debris consisting primarily of stones, scree, loam and sand; either formed by the accumulating effect of the ice and subsequently by the sea – slightly wavy, consisting predominantly of stones and scree (e.g., moraine coasts and firths), or sea-formed, slightly curved, smoothed-over course (e.g., shallow bay coast and sandbar/lagoon coast or marine-aeolically formed (by sea and wind), such as dune-wall coasts, for example (cf. op. cit.).

5.1.3 Sand Dunes

Due to the dynamics of sand, wind and water, coastal dunes are among the most diverse landscapes in the world. In a constant process, new windblown sand is deposited, alternating with valleys, lakes, marshes, brush or forests. Dune areas can be found in almost all coastal regions in Europe, sometimes dispersed in a rocky bay, sometimes over large vast areas of flat coastal landscapes. They contribute to protecting mudflats (wadden of the North Sea) or lagoons.

5.1.4 Tidal Coasts/Mudflats

Tidal coasts are formed by the constant rhythm of high and low tides: the twice-daily flooding and subsequent receding of water. A unique amphibian habitat is the Wadden Sea. It was formed in the postglacial period and consists of a sediment layer of sand and silt 10 to 20 m thick of varying composition. Extensive sand and silt areas alternate, forming a complicated fine relief which, due to the tidal streams, is subject to great dynamics (FAIRBRIDGE 1968).

Tidal coasts often have a very wide variety of flora and fauna. Together with salt marshes they constitute the biologically richest natural ecosystems. This is especially true of the Wadden Sea (cf. chapter 9.5, North Sea) and of the Coto Donana near Seville in southern Spain and the Ria Formosa in the south of Portugal.

5.1.5 Salt Marshes

Salt marshes are coastal areas formed from sedimentation of sand and silt mudflats and having saline plants (halophytes). Among these plants there are some which can live in salt water and can cope with periodic dry spells and others which are actually classified as terrestrial but which can cope with periodic flooding by seawater. Salt marshes lie at the upper limit of the tidal zone, so they are sure to be completely inundated by salt water during spring tides. In Europe salt marshes primarily occur on the Danish, German and Dutch coasts of the North Sea and in some parts of the north and west coasts of France and the east coast of England (GIERLOFF-EMDEN 1980).

5.1.6 Deltas

River deltas are at the borderline between salt and fresh water, and they, too, form complex, highly diversified landscapes. Away from the main currents little sand banks and small dunes can be formed. These in turn protect mudflats, salt meadows, salt marshes and lagoons. Typical examples in Europe are the Ebro delta in Spain, the Danube delta in Rumania and the Camargue at the mouth of the Rhône.

5.1.7 Estuaries

Low and high tides can shape river mouths into broad funnels called estuaries. In such estuaries, surrounded by mudflats and salt meadows, tidal changes are perceptible far into the interior. Fluctuations in the salinity of the water have an impact several kilometers up the rivers and produce a great diversity of life forms. One example is the Wester Schelde (Escaut) in the Netherlands.

5.1.8 Lagoons

If a bay or estuary is separated from the sea, for example by dune formation, a lagoon is the result. Lagoons are particularly prevalent in Great Britain, France and Spain. The size and depth of these coastal lakes and the salinity of the water can vary widely; at times the water is almost fresh. An example is the Albufera of Valencia, part of which is a nature reserve.

5.1.9 Marine Ecosystems

Marine systems consist of various moving water masses which can be distinguished according to their characteristic plankton composition. The sea-floor type is defined by the types of sediment it contains and the characteristic flora and fauna associated with it. In some regions, however, occasional or permanent lack of oxygen, as well as turbidity, amount of light or salinity can be the crucial factors in the formation of an ecosystem (MITCHELL 1987 in EEA 1995).

5.1.10 The Significance and Value of Marine and Coastal Ecosystems

The significance of coastal and marine ecosystems can be illustrated in exemplary fashion using the Wadden Sea: the Wadden Sea of the North Sea is the largest contiguous silt region in the world, with changing tides and, *in terms of biomass, it is as productive as a rain forest* (Common Waddensea Secretariat 1990 in EEA 1995). This high productivity is a result of the tides, influx of nutrients and a surplus of water.

Detritus and protective areas on the fringe of the marsh offer outstanding living conditions for young fish. Along with dunes, rocky coasts and lagoons, the large and often unspoiled wetlands form a habitat for millions of birds nesting and resting there. Coastal and marine ecosystems also fulfill a multitude of characteristic functions for coastal landscapes and for the people living there, for example:

- as natural protective systems against the flooding of the hinterland, shore erosion or storm damage;
- as natural purification systems by absorbing greater amounts of the nutrients and pollutants transported from the land into the coastal waters (e.g., "bodden" – shallow bays) and coastal ecosystems (e.g., salt meadows).

5.2 The European Seas in the Context of the World's Seas

The European seas, with their various subseas and bays *(Fig. 9)* are of great significance for maintaining the natural balance of the earth's biosphere. These include the Mediterranean Sea, Black Sea, Caspian Sea, White Sea, Barents Sea, Norwegian Sea, Baltic Sea, North Sea and the Atlantic Ocean.

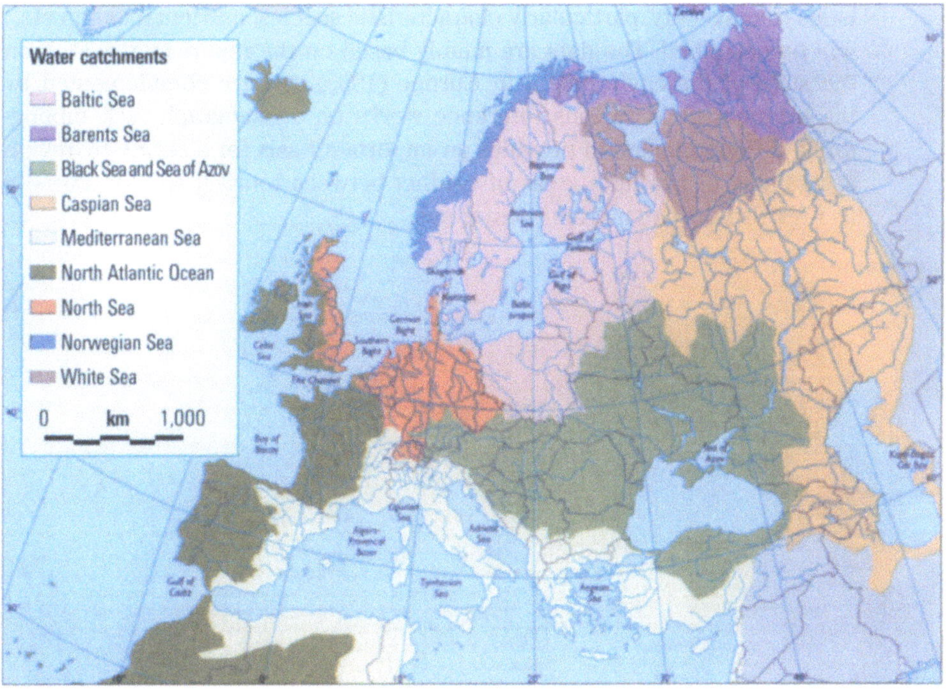

Fig. 9. The European seas and their catchment areas. Source: EEA 1995, p. 110, altered.

Distinction is made between open seas (Atlantic Ocean), semi-landlocked seas (North Sea) and landlocked seas (Black Sea). The quality of the semi-landlocked and landlocked seas and their biodiversity is influenced by a number of natural physical processes, such as alternating influx and loss of fresh water. When the evaporation rate is greater than the freshwater influx, the surface water becomes that much more dense, sinks more easily, leading to a strong vertical mixing of the water and thus to a high oxygen content in the deeper layers of the sea and in turn to a high diversity of biotic communities and great biodiversity. This is the case in the Mediterranean, for example. In the opposite case, i.e., when the evaporation rate is lower than the influx of fresh water, such as in the Black Sea, the oxygen does not reach the deeper layers of the sea, resulting in a lower diversity of biotic communities.

5.3 Individual Characteristics of the Seas

The following tables provide an overview of the major characteristics of the European seas. Following the basic data (geographic location etc.), the physical characteristics of the various seas and their biological characteristics (biodiversity, particularly characteristic species, nutrient makeup etc.) will be described. The data are mainly based on the report by the EEA on the state of the environment in Europe (1995) and are complemented by the information in various reference works on oceanography (cf. bibliography). The compilation provides an important basis for a region-by-region assessment of potential areas of conflict between tourism and biodiversity as regards Europe's marine ecosystems.

Mediterranean Sea

Basic data

- area: 2,516,000 km^2
- water volume: 3,758,000 km^3
- average depth: 1,500 m
- maximum depth: 5,092 m
- average salinity: 38–39.9 ppt
- average temperature: 12.7–14.5 °C
- annual average: n.d.
- January/June: n.d.

Major currents:
Water of the North Atlantic enters at the Strait of Gibraltar and follows the North African coast; some of the water circulates to the Ligurian Sea north of Corsica, other portions flow to the Tyrrhenian Sea. Moreover, North Atlantic water flows via the Strait of Sicily to the Eastern Mediterranean and from there spreads to the Adriatic and east as far as Cyprus.

Other facts:
- the largest of the semi land-locked European seas;
- bordered by 18 countries and three continents;
- two main basins, the Eastern and Western Mediterranean, and numerous smaller side arms;
- with 100 million tourists annually the coasts of the Mediterranrean are among the major tourist destinations in the world.

Physical characteristics

- Low precipitation,
- high evaporation,
- high salinity,
- little surf,
- relatively little nutrient concentration outside the inner coastal region;
- the coasts in the Northwestern Mediterranean basin are the most severely impacted by pollution; the concentration of population, industry and other factors is greatest here.

- Higher evaporation than precipitation; therefore potential of drying out, which is only prevented by the influx from the Atlantic.

Biological characteristics

- High species diversity,
- but low biological productivity due to low nutrient concentration;
- estimated total number of species: over 10,000;
- species diversity is higher in the Western than in the Eastern Mediterranean basin;
- there are a great many endemic species;
- especially the fauna is much more diverse than on the Atlantic coast;
- the most highly endangered species is the seal *Monachus monachus*, one of the ten most endangered mammals in the world;
- globally endangered species (IUCN) are the sea turtles *Caretta caretta* and *Chelonia mydas*; an average of 2,000 turtles lay their eggs on Mediterranean beaches each year, most of them in Greece and Turkey;
- approx. nine whale and dolphin species;
- vast beds of eelgrass *Posidonia oceanica*, which plays an important role in the ecosystem: production of organic material, primary producer of oxygen, habitat of many fish species, halts sedimentation, protects beaches by braking surf;
- important fish stock, which, however, is severely decimated in the Northwestern Mediterranean.

Table 13.1. Mediterranean Sea.

Black Sea (B) *with Sea of Azov (A)*

Basic data

coastline: n.d.

- area (B) 461,000 km²
  (A) 38,000 km²
- water volume
  (B) 537,000 km³
  (A) 300 km³
- average depth
  (B) 1,166 m
  (A) 8 m
- maximum depth
  (B) 2,245 m
  (A) 13.5 m
- average salinity:
  (B+A) 16–18 ppt
  (at depths of 150–200 m: 21–22.5 ppt)
- average temperature: n.d.
- annual average: n.d.
- January–June
  summer: 25–28 °C
  winter: 6–8 °C

Major currents

two cyclonic currents moving counterclockwise, one in the western and one in the eastern basin

Other facts

- world's largest landlocked sea
- bordered by six countries
- the northwestern basin is most heavily impacted by anthropogenic influences
- 40 million tourists in the summer months on the coasts of the Black Sea

Physical characteristics

- The connection of the Black Sea to the Mediterranean is shallow (50 m) and narrow
- more water flows from the Black Sea into the Mediterranean than vice versa; the deficit is made up for by numerous freshwater sources, most of which, however, are enriched with too many nutrients today
- characteristic for the Black Sea is an extremely high salinity and a strong vertical layer structure of water masses of varying density
- 90% of all the water is without oxygen
- accumulation of large quantities of organic material which is broken down at greater depths; in the process the oxygen present is converted to nitrate and sulfate compounds.
- severe pollution from discharges from various countries, combined with low vertical mixing of the sea, greatly aggravate the natural oxygen deficiency
- river regulation and increased dam construction have lessened freshwater influx, resulting in a higher sedimentation and severe coastal erosion

Biological characteristics

- Due to the low oxygen content and the high salinity there are comparatively few species in the Black Sea
- 180 different fish species; decline of large fish species in favour of small ones (cf. BRONFMANN 1993).
- three dolphin species
- in the protected areas of the Danube delta alone 92 different fish species. The region is also very significant ornithologically: 225 species of water birds live here; seabirds use the region as a nesting and rest area
- The Black Sea is impacted by various severe ecological problems. In the last 30 years it has undergone a transformation from a "healthy" ecosystem to a hypereutrophic system which has become uninhabitable for most higher organisms. The population of Black Sea dolphins has been reduced to one tenth; many of the animals suffer from skin diseases (op. cit.)
- This is further aggravated by the fact that in the meantime a few "opportunistic" species have found niches in the Black Sea, and this has contributed to crowding out other originally endemic species

Table 13.2. Black Sea with Sea of Azov
Table 13.3. *(right)* Caspian Sea.

Caspian Sea

Basic data

- coastline: 1,200 km
- area: 421,800 km²
- water volume: 77,000 km³
- average depth: 76 m
- maximum depth: 5,092 m
- average salinity varies from region to region:
  0.3 ppt (North Caspian Sea)
  14 ppt (southeast)
  300 ppt (Gulf of Kara Bogaz)
- average temperature: n.d.
- annual average: n.d.
- January/June:
  summer: 22 °C
  (east coast: 27–30 °C),
  winter: in part frozen over

Major currents:

cyclonic water circulation with two main currents in the central and southern basin; also, coastal current along the west coast from the Volga delta towards the south and from the east coast to the north

Other facts:

- the world's largest landlocked sea of low salinity,
- situated in an arid and semiarid zone, bordered by 5 countries,
- important source of petroleum and natural gas, many drilling rigs

Physical characteristics

- Physically, the Caspian Sea is divided into the northern, central and southern sea.
- The salinity of the water increases from north to south.
- Three quarters of the freshwater influx stems from the Volga.
- Other rivers emptying into the Caspian Sea are: the Ural, the Kura, the Kuma, the Terek and the Samur.
- In the eastern part of the sea there is no significant influx, so with high evaporation the salinity is particularly high; in this region various minerals are mined.
- Vertical mixing occurs only to a very limited extent, leading to low oxygen content in depths of 200 to 300 m. This in turn results in increased production of hydrogen sulfide.
- The Caspian Sea has risen markedly, which presumably has to do with the hydroengineering projects in and along the Volga bed. In the north of the region this rise in sealevel has caused great problems in coastal settlements and on beaches.
- The surface water temperature is between 24 und 27 °C in the summer; in the winter it sinks in the south to 9 °C, in the north to 0 °C.
- The Volga is the European river with the largest catchment area: 15 large cities are situated on its tributaries and branches and a large number of large industrial sites. The Volga is for this reason not only the greatest source of water for the Caspian Sea, but also the leading source of danger due to the discharge of pollutants into its northern portion.

Biological characteristics

- The lack of oxygen in the deeper layers has resulted in a dearth of sea fauna in the Caspian Sea, particularly in the central and southern portion.
- The biological diversity of phytoplankton, zooplankton and larger marine animals is steadily decreasing.
- Particularly affected are the central and southern Caspian Sea. In 1988–89 hundreds of thousands of dead fish (mainly sturgeons) were discovered here, many of them with noticeable physical changes indicating severe poisoning (cf. EEA, p. 127).

White Sea

Basic data

- coastline: n.d.
- area: 95,000 km
- water volume: n.d.
- average depth: 100–200 m
- maximum depth: 340 m
- average salinity:
  24–26 ppt (surface water)
  30–30.5 ppt (deep water)
- average temperature: n.d.
- annual average: n.d.
- January–June: n.d.

Major currents

freshwater influx, discharge into Barents Sea

Other facts

- semi-enclosed sea of the Arctic Ocean situated on the northern coast of Russia.
- three major rivers empty into the White Sea: Svernaya Dvina, Mezen and Onega, all three with very large catchment areas.

Physical characteristics

- Bays freeze over in winter
- tides up to 7 m.
- the central portion is separated from the Barents Sea by a shallow reef, so that the deep water is not mixed

Biological characteristics

- The formerly rich fish stock of the White Sea, the numerous seal banks and vast eelgrass areas have been severely decimated in recent years. The reasons for this are above all the high urbanisation and environmental pollution (particularly material discharge via the three major rivers).
- The eelgrass areas are estimated to have decreased by half between 1975 and 1989.
- The population of Greenland seals is also said to have been reduced to half in only four years (1985–1989).

Table 13.4. White Sea.

Barents Sea

Basic data

- coastline: n.d.
- area: 1,425,000 km²
- water volume: 322,000 km³
- average depth: 299 m
- maximum depth: 600 m
- average salinity: 35 ppt
- average temperature: 4–12 °C
- annual average: n.d.
- January/June: n.d.

Major currents:
Three currents, three different masses of water, are mixed in the Barents Sea and determine its physical state: the Norwegian current with relatively low salinity and high temperatures, the Atlantic current with high salinity and temperatures between 3.5 und 6 °C and the very cold Arctic current from the north.

Other facts:

- on the fringe of the Atlantic Ocean, situated at the outermost edge of the Greenland Sea;
- the coastline is strongly characterised by fjords and gulfs;
- the most important river emptying into the Barents Sea is the Pechora;
- the coastal region is scarcely settled; the major form of exploitation is fishery;
- increasingly frequented by cruises.

Physical characteristics

- The Barents Sea is primarily characterised by the exchange of warm water masses of the Atlantic in the south with the cold, Arctic water masses in the north;
- during winter a strong vertical mixing of the water masses occurs which is of vital significance for primary production in spring and summer;
- hardly any sediment-bearing rivers empty into the Barents Sea, which is consequently unusually clear and transparent;
- very stormy body of water with unstable weather conditions.

Biological characteristics

- The Barents Sea is an Arctic sea of high biological production.
- During the six summer months the Barents Sea provides nearly optimum living conditions, with intensive insolation and sufficient nutrient supply; according to the EEA, the area of the polar cap is one of the most significant regions of biological production.
- The high phytoplankton and zooplankton content is the basis of the large fish stock in the Barents Sea;
- there are large seabird populations here;
- the region is moreover very significant in regard to sea mammals.
- Because the biological production is very limited in time frame and space, the ecosystem is potentially very sensitive to anthropogenic influences and impairments.
- Apart from the large fish stock, there are also various "itinerant" species of whale and seals and polar bears in the Barents Sea.
- Beginning in late autumn nearly the entire Barents Sea is covered with ice. During the summer thaw, large amounts of drift-ice floats along the shore; as a result, only a few plant and animal communities establish themselves near the shore. At depths of from 5–10 m are brown and red algae.

Table 13.5. Barents Sea.

Norwegian Sea

Basic data

- coastline: n.d.
- area: 1,383,000 km²
- water volume: 2,408,000 km³
- average depth: 1,742 m
- maximum depth: 3,960 m
- average salinity: 34.9–35.4 ppt
- average temperature
  100–600 m (depth): 9 °C
  > 600 m (depth): 1 °C
- annual average: n.d.
- January/June: n.d.

Major currents:
The Norwegian Sea consists of four different, constantly mixing water masses: the warm, salty water of the Atlantic (temperature > 8 °C, salinity 35/1000), the cold Arctic water (< 3 °C, salinity between 34.7 and 34.9/1000), moreover, ground water from the Greenland Sea and the Arctic Ocean, which fills the lower layers and makes up the greatest volume. The coastal water from the Norwegian coast is the fourth major current: a mixture of fresh water from the land and water of the Atlantic Ocean, resulting in relatively low salinity.

Physical characteristics

- The Norwegian Sea is largely free of ice even in winter

Biological characteristics

- In the spring, the water of the Norwegian Sea is extremely rich in nutrients, offering great potential for primary production, particularly in shallow coastal areas
- this results in there being great stocks of fish in the Norwegian sea, which in turn serve as food for seals and whales
- the whale stock has been estimated at 10,000 during the summer months

Other facts:

- comprises two basins separated by a mountain range between the Voring Plateau and Jan Mayen;
- the coastal regions of the Norwegian Sea are very sparsely populated
- The main forms of exploitation of the coastal and marine areas are fishery, mining and a few heavy industries; the sea is also strongly frequented by cargo ships and contains petroleum and natural gas deposits
- by this time also strongly frequented as a tourist destination for cruises etc. (cf. Barents Sea)

Table 13.6. Norwegian Sea.

Baltic Sea

Basic data

- coastline: 8,045 km
- area: 386,000 km^2
- water volume: 33,000 km^3
- average depth: 86 m
- maximum depth: 459 m
- average salinity: 0–25 ppt, according to region
- average temperature: surface water: 0–18 C, deep water: 2–5 C
- annual average: n.d.
- January/June: n.d.

Major currents:
Primarily determined by freshwater influx from the surrounding land masses. The Baltic is fed by many rivers. The Vistula and Oder are the largest of them and make up 23% of the total freshwater influx. Water entering from the North Sea spreads northward counterclockwise along the German Baltic coast to the west coast of Finland and then south along the east coast of Sweden back towards Denmark.

Other facts:

- a shallow, relatively young sea formed in the last Ice Age, consisting of many basins.
- the coastal area of the Baltic is densely populated (77 million people) and highly industrialised; moreover, large areas are devoted to agriculture.

Physical characteristics

- The salinity of the Baltic varies according to region: in the Kattegat it is between 15 and 30 ppt, in the open Baltic between 6 and 8 ppt, in the Gulf of Finland only 2 to 7 ppt.
- By nature, the oxygen content decreases with the depth of the water layers; however, on the floor of the Baltic there is virtually no oxygen at all. This is primarily due to lack of vertical mixing of the water layers and the oxidation of organic material.
- The nutrient content of the Baltic is steadily increasing; the production of phytoplankton appears to have nearly doubled in the last 25 years, severely reducing oxygen content.
- Little mixture of the water masses;
- the northern portion of the Baltic is often covered with ice over 6 months of the year, so that biological production occurs only during 4 to 5 months, compared to the 8–9 months in the southern regions.
- Due to the relatively low water volume and low water exchange (the water of the Baltic stays in the Baltic approximately 30 years), the Baltic is very sensitive to all kinds of pollution, in particular to nutrient enrichment.

Biological characteristics

- The Baltic is very deficient in species, both flora and fauna.
- Most of the species migrated there from neighbouring regions: primarily marine organisms from estuaries and shallow coastal areas of the North Sea and relics of the Ice Age. The number of marine species is markedly decreasing from the Danish Basin to the Gulf of Finland. Of the 1,500 invertebrate species of the Norwegian coast, only 70 are found in the Baltic, and of the many hundreds of marine algae species on the Norwegian coast, only 24 are now found on the Finnish coast. Various organisms from the fresh water from rivers and lakes or from the sea are found in the Baltic in morphologically altered states.
- Among the most endangered species of the Baltic are all the indigenous seal species, whose population has been strongly decimated in recent decades. The numbers of sea urchin have also markedly decreased. The reasons for this are primarily the pollution of the Baltic (the discharge of toxic matter has drastically increased), but also hunting and modern fishery practices.

Table 13.7. Baltic Sea.

North Sea

Basic data

- coastline: 6,432 km
- area: 750,000 km²
- water volume: 94,000 km³
- average depth: 94 m (in the Norwegian Basin)
- maximum depth: 700 m
  average salinity: 31–35.25 ppt
  average temperature: 9.5–11.5 °C
  annual average: 9.5–11.5 °C
  February/August: 2–7 °C/11–17 °C

Major currents

Water from the Atlantic flows in from the north. 90 % of the inflow takes place between Scotland and the Orkney Islands and stems from the North Atlantic current. The current flows counterclockwise to the south along the Norwegian coast, leaving the North Sea as a Baltic current.

Other facts

The catchment area of the rivers emptying into the North Sea is 850,000 km². The catchment areas of the Elbe, Weser, Rhine, Thames, Humber and Seine are densely populated (165 million people), highly industrialised and intensively cultivated. Via these rivers correspondingly large amounts of nutrients and pollutants are discharged into the North Sea.

Physical characteristics

- Twice daily the tide inundates the wadden of the North Sea coast, covering sand and silt, eelgrass meadows and clam banks for kilometers.
- The salinity of the water fluctuates constantly: at rainfall the water becomes as brackish as in river mouths.
- At low tide, flora and fauna are exposed to the "whimsy" of land weather: scorching heat in the summer with water temperatures over 30 °C in shallow swamps, and frost and drift ice in winter.
- Most regions of the North Sea are vertically well mixed in winter; from springtime on a thermal boundary comes about in many areas separating the layers from each other. An exception is a coastal strip in the southern North Sea in which vertical mixing takes place throughout the year (from northern France to the German basin).
- The climate of the North Sea is primarily determined by the Atlantic: many winds from varying directions and at varying strengths, high cloud density and a relatively high precipitation rate (averaging 425 mm/a).

Biological characteristics

- Many of the invertebrates hidden in the sand or silt are ecological generalists which tolerate constant fluctuations. Other animals avoid temporary worsening of their habitat by long distance migration, or they produce great amounts of progeny to make up for losses.
- At high tide flounders, soles, herring, sand shrimps and many other species profit from the abundance of nutrients in the wadden sea. When the water recedes, they retreat to the narrow channels again, leaving the spaces now drying up to the birds.
- During the summer months, primarily waders, terns and gulls nesting in the foreshore areas look for food here. In spring and autumn, millions of migratory birds pause in the wadden sea to take on energy reserves for their continued journey.
- Seals congregate at low tide in their traditional spots on sand banks or beaches. Here they also rear their young and molt.
- Various dolphin species and a few whales also have their habitat in the North Sea during some months of the year.
- Since 1600, due to dike construction, 80 % of the salt meadow areas of the foreshore have disappeared.
- From all sides great amounts of pollutants and nutrients are being discharged into the North Sea: mercury, lead, cadmium, chlorohydrocarbons, nitrogen and phosphorus.

6 European Biodiversity

In Europe, biodiversity is very unevenly distributed, with the least variety of ecosystems, accompanied by the lowest species diversity, in the north of Europe (EEA 1995). Centres of high biodiversity are to be found in the Mediterranean region (Italy, Spain, Greece, France) and on the fringes of Europe (Bulgaria, Ukraine, Georgia, Armenia), with over 5,000 endemic plant species which occur only in these countries. (According to the EEA, many of the European countries list species as endemic, although the same species might also exist in another country but is not [yet] listed and protected there – a consequence of differing methods of data compilation. There is a lack of internationally recognised regulations to overcome this data deficit and ensure comparability.)

Other "indigenous" European plants, in turn, were originally brought there from other countries, cultivated and are now regarded today as indigenous or even endemic. Their genetic material is usually still to be found in their countries of origin, and vice versa. Spain, Italy, Greece and the former Yugoslavia have the highest incidence of indigenous plant species and Ireland and the Faeroe Islands the lowest *(cf. Fig. 10)*.

Table 13.8. *(left)* North Sea.

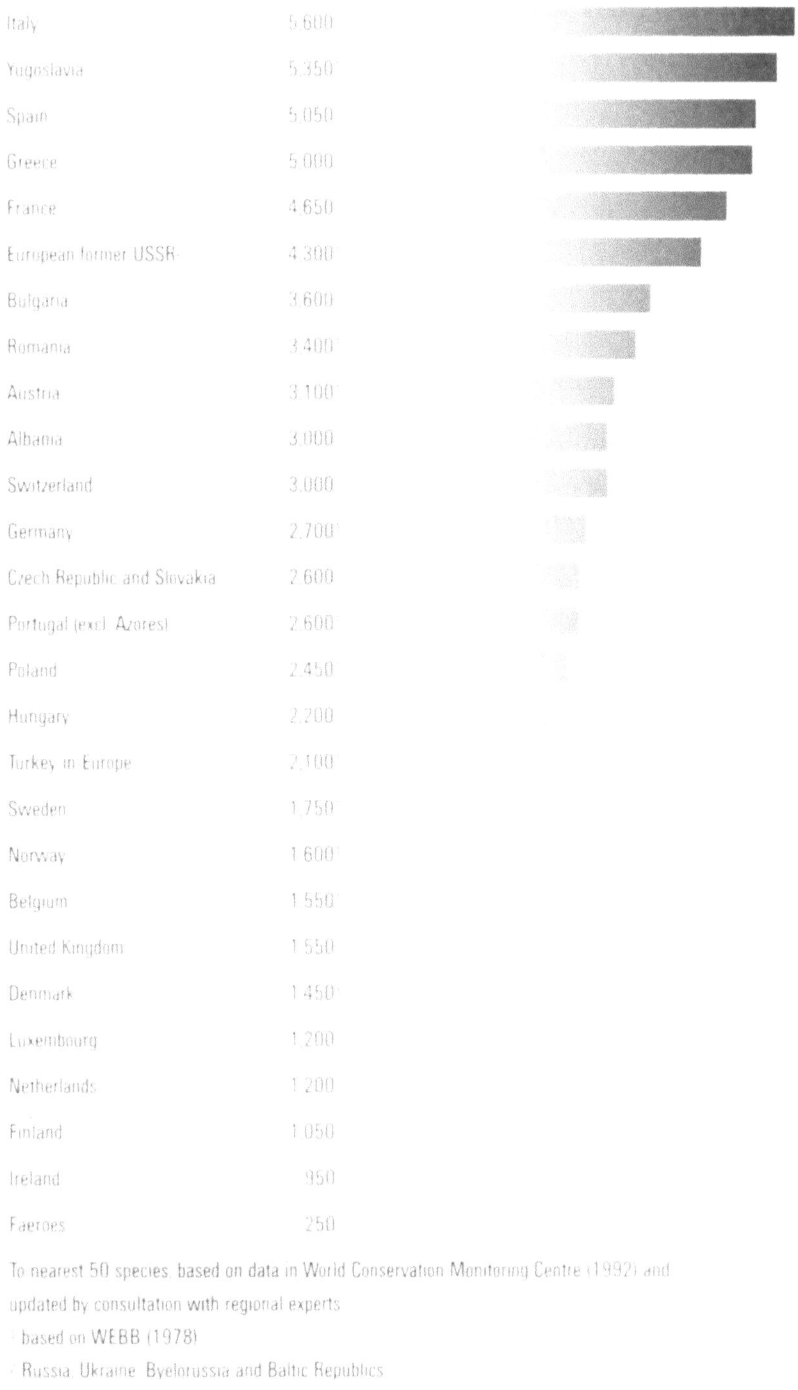

Fig. 10. Number of indigenous plant species in individual European countries. Source: IUCN 1994, p. 42, altered.

6.1 Regions of High Significance for Biodiversity

To assess the significance of the various countries of Europe in terms of biodiversity, species diversity and density and the degree of endemicity are evaluated quantitatively below, ranked accordingly and combined, in analogy to the overviews of the IUCN (1994) and WCMC (1994). As the "species-diversity" factor predominates in the quantitative surveys of the IUCN and WCMC, the focus herein will be on significance for species diversity. The result is a ranking according to the categories low, medium, high or very high, based on total land area *(cf. Table 14 and Map 6).*

Europe/country	Number of species	Species density	Endemic species	Biodiversity
Albania	high	high	medium	high
Belgium	medium	medium	low	medium
Denmark	medium	medium	low	medium
Faeroe Islands	low	medium	low	low
Germany	high	low	medium	medium
Estonia	medium	medium	low	medium
Finland	low	low	low	low
France	very high	medium	medium	high
Gibraltar	low	very high	low	medium
Greece	very high	medium	high	very high
Ireland	low	medium	low	low
Italy	very high	medium	high	[illegible]
Latvia	low	medium	low	low
Lithuania	low	medium	low	low
Malta	low	high	low	medium
Netherlands	medium	medium	low	medium
Norway	medium	low	low	medium
Poland	medium	medium	low	medium
Portugal	high	medium	medium	medium
Spain	very high	medium	high	very high
Sweden	medium	low	low	medium
Great Britain	medium	low	medium	medium
Yugoslavia (former)	high	medium	medium	high
Turkey (Asia)	very high	medium	very high	very high

Table 14. Assessment of biodiversity in relation to total land area. Source: calculations and assessments based on WCMC data.

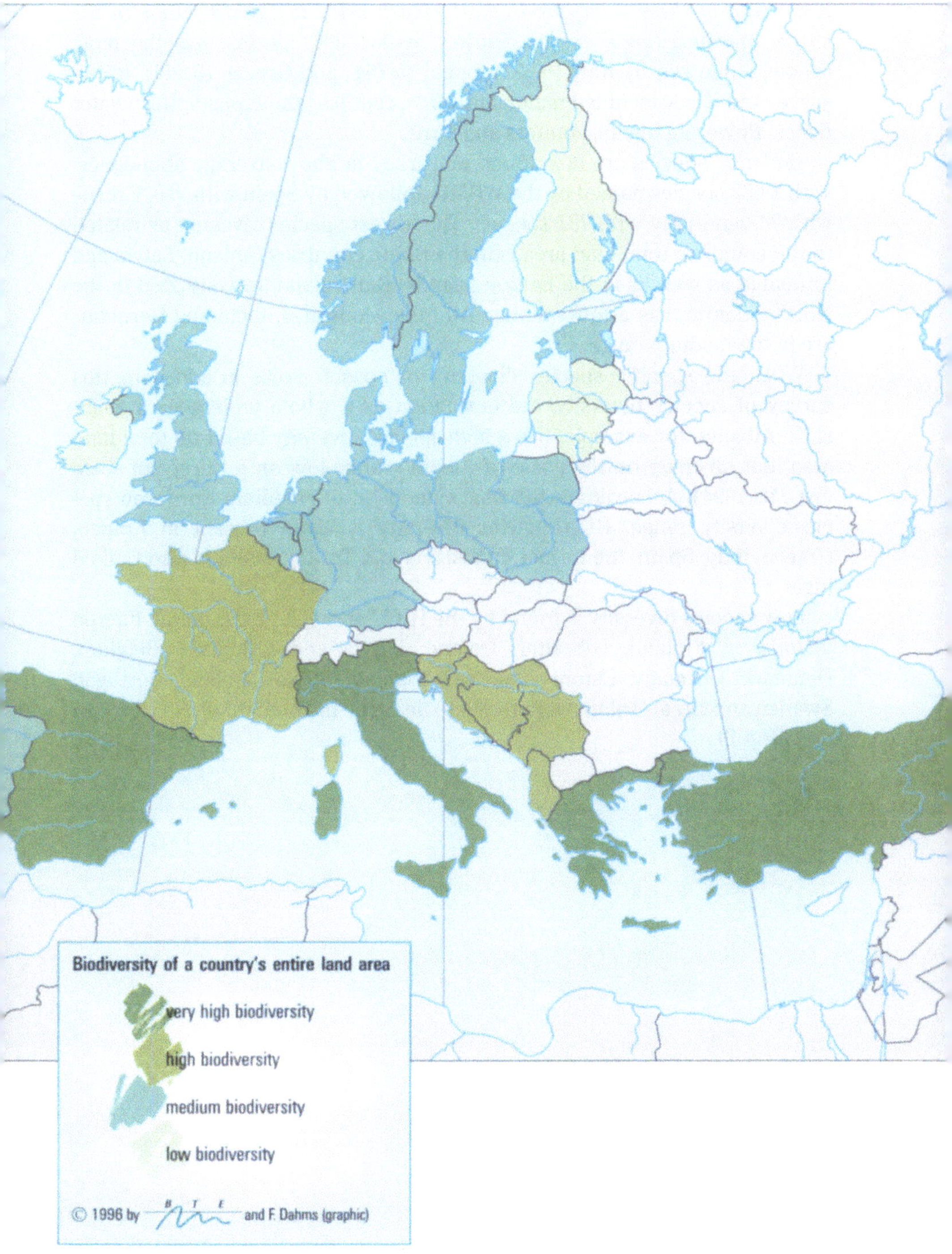

Map 6: Quantification of biodiversity with regard to total land area.

According to these data, the highest biodiversity is clearly found in the entire Mediterranean region. Turkey, with 9,383 species, has the most species, followed by Italy (6,252), Spain (5,764) and Greece (5,649). In the survey the IUCN included mammals, birds, reptiles, amphibians, freshwater fishes, flowering plants, conifers and ferns.

Endemic species are also found in Turkey in above-average abundance, with 2,682 species named by the WCMC, followed by Spain with 960, Greece with 748 and Italy with 721 species. The lowest species diversity as related to the country's total land area is in the Baltic countries Finland, Latvia and Lithuania as well as in the Faeroe Islands. But Ireland, too, situated in the North Atlantic, has a low ranking. All other countries, including Germany, are in the medium range.

When the specific species diversity of coastal areas is added to this survey of species diversity, the evaluation as a whole undergoes a slight shift. Albania, for example, has a high species diversity based on total land area, but diversity on the coast of Albania ranks low on a European scale (WCMC 1994). Accordingly, the final value is in the medium European species-diversity range. High species diversity is again present in France, Greece, Italy, Spain, the former Yugoslavia and Turkey. Malta is also ranked high.

Low species diversity is found by the IUCN survey to exist on the Faeroe Islands, in Finland, Gibraltar, Ireland, Latvia and Lithuania. Belgium, Denmark, Germany, Estonia, the Netherlands, Norway, Poland, Portugal, Sweden and Great Britain were found to lie in the medium range *(cf. Table 15 and Map 7).*

Europe/country	Species diversity in relation to land area [1]	Species diversity on coast [2]	Biodiversity *total* [3]
Albania	high	low	medium
Belgium	medium	low	medium
Denmark	medium	medium	medium
Faeroe Islands	low	low	low
Germany	medium	medium	medium
Estonia	medium	low	medium
Finland	low	medium	low
France	high	medium	high
Gibraltar	medium	low	low
Greece	very high	medium	high
Ireland	low	low	low
Italy	very high	medium	high
Latvia	low	low	low
Lithuania	low	low	low
Malta	medium	very high	high
Netherlands	medium	medium	medium
Norway	medium	high	medium
Poland	medium	niedrig	medium
Portugal	medium	medium	medium
Spain	very high	medium	high
Sweden	medium	medium	medium
Great Britain	medium	medium	medium
Yugoslavia (former)	high	medium	high
Turkey (Asia)	very high	medium	high

Table 15. Grades of biodiversity in European coastal areas.

[1] species diversity + species density + number of endemic species

[2] species diversity only

[3] species diversity of country + species diversity on coast

Source: calculations and rankings based on WCMC data, 1994.

Map 7. Quantification of biodiversity in European coastal areas.

6.2 Endangered Biological Diversity in Europe

Worldwide it is a declared goal of nature conservation to protect species diversity. However, species diversity as determined by the number of species is not an adequate indicator for the protection of biotic communities. What is more urgently needed for the preservation of biodiversity is access to precise information about the species community typical for the nature area and habitat. This is only possible when detailed distinctions are arrived at as to which spatial structures and biotic communities are specific for a landscape, i.e., when registers of nature areas have been compiled (cf. BLAB et al. 1995). This demand has indeed been recognised throughout Europe, as shown by the FFH Directive of the European Union (cf. chapter 10), but implementation at the national level has been of widely varied stringency and will be long in offering comparable data and points of departure. All the significant species and biotic communities of Europe are far from being identified and classified (cf. EEA 1995). Therefore focus will be concentrated herein on the *degree of endangerment of the species*, which, thanks to monitoring systems long since established by scientific institutions and individual experts, already affords a quite detailed picture of the European region. *Table 16* shows the percentages of species threatened by extinction worldwide and in Europe. For lack of data entire groups of species are not included, such as algae, bacteria, fungi, lichens etc., so that one must assume that even more species are threatened by extinction.

	World		Europe	
	Total number	Threatened (%)[1]	Total number	Threatened (%)
Mammals	4,327	16	250	42
Birds	9,672	11	520	15
Reptiles	6,550[1]	3	199	45
Amphibians	4,000	2	71	30
Fish (freshwater)	8,400[1]	4	227	52
Invertebrates	>1,000,000[1]	?	200,000[1]	?
Higher plants	250,000[1]	7	12,500[1]	21

[1] estimates

Table 16. Number of species in danger of extinction worldwide and in Europe. Source: EEA 1995, p. 222, altered.

In Europe 52% of all freshwater fish, 42% of the mammals and 45% of the reptiles are threatened by extinction. As to higher plant species, 21% are threatened.

6.2.1 Endangered Flora

Compared to other regions of the world, Europe has relatively few plant communities and species – a consequence of the ice ages. 12,500 plant species are (unevenly) distributed throughout Europe. In surveys, all the coastal areas have not yet been charted as ecosystems in their own right and treated separately from other habitats (cf. EEA 1995). The greatest species diversity is found in the Mediterranean region, as shown in chapter 6.1. The greatest share of endemic plant species has been found to exist in the mountains of Europe: in the Alps, the Pyrenees and the Sierra Nevada. The Sierra Nevada alone is home to two thirds of all the endemic plants in Spain. The *Canary Islands*, with a share of *48%*, has a rate of endemicity far above average; that of *Corsica* is *38%*.

Particularly endangered are the relatively few plant communities in Central and Atlantic Europe. The main problem in these regions is the dramatic drop in species diversity found in nearly all ecosystems. The lower the diversity of species, the more sensitive they are to disturbances. Central Europe, with between 25 and 35 per cent, has the largest share of endangered plant species in Europe.

A great many of the plant species occur exclusively on coasts. On dune coasts these are, for example, beach grass, sea holly and scurvy grass. On rocky coasts, in turn, many specific algae, moss and lichen species grow, as well as other characteristic plants such as sea fennel (cf. EUCC 1992).

Primarily in the coastal areas of the Mediterranean, tourism is considered to be the main source of disturbances and destruction of the plant world typical of the coast and sea (cf. chapter 8).

In Northern Europe there are comparatively few endangered species.

All in all, 27 higher plant species of Europe are said to have died out; 2,200 have been rated by the IUCN as "endangered" (in various degrees of endangerment; cf. EEA 1995).

6.2.2 Endangered Fauna

Europe is the habitat of 250 different mammalian species, 520 bird species, 199 reptilian species, 71 amphibian species and approx. 200,000 invertebrates (estimates of the WCMC, 1992). The marine and coastal ecosystems fulfill vital functions for many species. *Cliffs and rocky coasts*, for example, are indispensable for many bird species for nesting (e.g., for the populations of gannets, puffins and also various sea-gull species, guillemots, falcons and storm petrels). The *beaches* and sandbanks afford habitats for seals and sea turtles, for example, but also for birds such as the sandwich tern. The large *dune* areas are in turn preferred nesting spots for dozens of bird species such as the wheatear and various duck species. Many other species use them as resting places during migration. *Mudflats* and *salt meadows* are important for waders such as oysterbirds, avocets and plovers. *Lagoons* and *coastal swamps* shelter vast numbers of pelicans and flamingos – in Europe, too. Wide *river mouths* are indispensable nesting places, e.g., for marbled ducks and stilts (cf. EUCC 1992).

The table below provides an overview of the most endangered species of the various families and the major sources of threat (not included are the fish stock in the European seas, as comparable data were only available for freshwater fishes). Only rarely are the sources of threat anthropogenic intrusions in the habitat alone, as generally several factors acting in parallel, temporally and spatially, impact the species and their habitats together. More precise studies of larger regions of Europe dealing with the direct impact of tourism on species and habitats are not available to date. An exception is the Blue Plan (1989), which provides an overview of the Mediterranean countries and which will be discussed in chapters 8 and 10 of this study. Otherwise only studies of individual cases (cf. chapter 9) in these countries enable an assessment of the potential danger emanating directly from tourism to be made. The clear results of these individual case studies permit the assumption to be voiced here that *tourism, when present at the site in question, is at least an indirect source of threat to all the species named*, as it is precisely construction and infrastructure steps taken by touristic projects with which disruptions and in many places destruction of habitats are associated (such as, for example, on the Costa del Sol in Spain). The effects of construction and tourist activities on coasts are particularly drastic on bird populations, reptiles and amphibians, as shown in exemplary fashion by the results achieved thus far by the "Ecosystem Research of the Schleswig-Holstein Wadden Sea" (1994; see also chapter 8). But certain mammalian species are also directly threatened by touristic activities, which is particularly clearly illustrated by the seal population in the Mediterranean (cf. EEA 1995), but also in the Wadden Sea (cf. ÖKOSYSTEMFORSCHUNG WATTENMEER 1994).

Mammals (M)

Total:
91% of **M** are indigenous species, *9%* introduced. *18%* of the indigenous species are endemic. 24 of the 698 **M** ranked worldwide as severely endangered are found in Europe, which amounts to *11%*.

Sea mammals:
Among the **SM** the threat is particularly apparent in *whales*. Here the conflict is primarily between the interests of nature conservation and the business interests of the international whaling organisations.

Another example is the seal *Monachus monachus* indigenous to the Mediterranean which the IUCN ranks amongst the world's 10 most endangered **M**. Today there are only some 500 individuals left, and their numbers continue to decline. The main influences on the seal population are hunting, coastal construction projects for facilities of Mediterranean tourism and severe water pollution.

Birds (B)

15% of all bird species in Europe are considered highly endangered. In Italy (*43%*) and France (*37%*), in the Netherlands (*35%*) and Germany (*40%*) their habitats as well as nesting and resting areas are considered the most endangered.
In wetlands and coastal regions, the decline of bird species is particularly great.

Reasons for impairments lie mainly in the *loss of undisturbed habitats due to tourism on the coasts* and bird hunting. Furthermore, the loss of feeding areas due to the use of pesticides is a major reason for species decline.

Reptiles (R) and amphibia (A)

The greatest number of **R** and **A** is in the Mediterranean and Southeastern Europe. **A** and **R** are extensively affected by continuing habitat destruction. *45%* of the **R** and *30%* of the **A** in Europe are regarded as endangered, including *sea turtles*. Nearly sedentary species combined with the necessity of wintering make **A** and **R** particularly sensitive even to temporary changes in their habitats.

*In the Mediterranean the increase in tourism in recent decades is primarily responsible for the decline of **A** and **R**.* Turtle nests, for example, are a touristic attraction. Sand dunes are important habitats for numerous reptiles, but are also increasingly being destroyed by the influence of tourism.

In other regions the influence of urbanisation, industry, road construction etc. are more severe factors.

Invertebrates (I)

Of the estimated *200 000* invertebrate species in Europe, *about half find their habitats in/on seas and coasts*. **A** precise, comprehensive overview of the potential distribution of **I** has yet to be compiled. Only a few national parks, such as the Berchtesgaden National Park, have compiled data of this nature, as have a few nature reserves. One of the most important functions of **I** is pollination. Another is the composting of natural wastes, soil aeration etc. Invertebrates are for most **R** and **A** a direct source of nourishment. 90% of European fishes and birds (as well as some mammals) also depend on invertebrates for nourishment.

On the degree of endangerment of invertebrates there is as yet no compilation of representative Europe-wide data. There are individual surveys on land invertebrates in various countries. According to them it is estimated that *between 15 and 40% of **I** should be ranked as endangered* in protective lists. On marine invertebrates there exist today no comparative data at all which could justify such estimates.

6.2.3 Endangered Ecosystems

Natural and relatively untouched natural ecosystems are today threatened by anthopogenic factors in many areas, particularly in Central Europe. Comprehensive data on the current state and the distribution of ecosystems do not yet exist for Europe as a whole (EEA 1995, p. 191).

In Germany the decrease in natural and nearly untouched natural habitats has led in the past decades to the fact that today some 110 natural ecosystem types with a total of 73,000 animal and plant species are limited to 3 to 5% of the country's area. All in all, the seas and coasts in Germany are endangered to an above-average degree (BfN 1995, p. 31). In the coastal regions the habitat types dunes, salt marshes and rocky coasts have undergone a dramatic decline in recent decades *(cf. Fig. 11)*. Tourist activities are considered to be one of the major causes for this in many regions of Europe. According to a study by RATHS et al. (1995), the main sources of threat to the ecosystems of the sea and coasts in Germany are:

1. soil, air and water pollution
2. soil and water eutrophication
3. mechanical influences
4. removal/stocking of plants and animals
5. expansion of coastal protection/waterways
6. intensification of use of coastal areas.

Tourism is a "coperpetrator" of these threats whose degree of impact depends on the intensity of the form of tourism in question and the conditions of the nature area. Particularly in the Mediterranean region, the impact of tourism in coastal areas is a major factor in the destruction and impairment of marine and coastal ecosystems (see also chapters 7 and 8). This is also illustrated by the rapid decline of species and habitats as shown in the following figure *(Fig. 11)*.

Table 17. *(left)* Fauna endangered by tourism. Calculations based on EEA data, 1995.

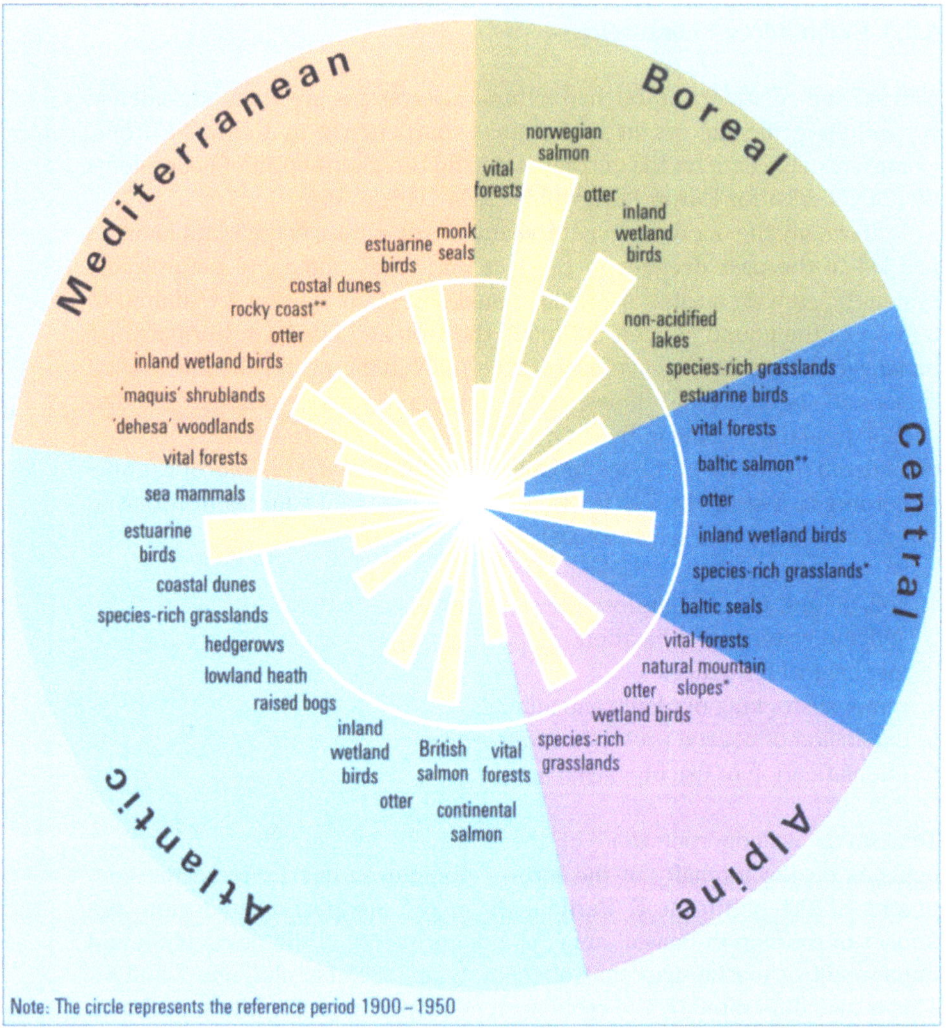

Fig. 11. Changes in population and habitat in various geographical regions from 1900 to 1950 (white circle denotes the year 1900). Source: EEA 1995, p. 222, altered.

Particularly drastic changes are shown by the decline of undisturbed rocky coasts and coastal dunes in the Mediterranean and by the decline of the monk-seal population. But other regions of Europe also display marked changes and losses. In the Atlantic region there is a decline of coastal dunes as well as sea mammals, thus illustrating the severe impairments of habitats and species. Only in parts of the boreal region have positive changes taken place, such as the increase in bird populations in wetlands in the continental interior.

6.3 Species and Habitat Protection in Europe

All European countries have issued guidelines for the protection of certain valuable areas from the standpoint of nature and landscape protection (cf. chapter 10). The IUCN (1993) defines a protected area as follows:

> *an area of land or sea especially dedicated to the protection of biological diversity, and of natural and associated cultural resources, and managed through legal or other effective means.*

The terminology denoting the various protected areas in Europe has thus far not been unified. On the contrary, there is a broad spectrum of names in the individual countries: biosphere preserves, forest preserves, sea preserves, nature parks, regional nature parks, landscape-protection areas, game preserves and many others. This is further complicated by the fact that the same terms denote different protection goals in different countries, and this leads to confusion at the international level. For this reason the IUCN has defined six different, internationally recognised protected-area categories, most recently in 1994, and laid down the criteria regarding their scope and protection goals in the various categories (cf. FNNPE 1993; EEA 1995):

IUCN Protected Area Categories

I	Strict Nature Reserve / Wilderness Area:	managed mainly for science or wilderness protection
II	National Park:	managed mainly for ecosystem protection and recreation
III	Natural Monument / Natural Landmark:	managed mainly for conservation of a specific natural feature
IV	Habitat and Species Management Area:	mainly for conservation through management intervention
V	Protected Landscape / Seascape:	managed mainly for landscape/seascape protection and recreation
VI	Managed Resource Protected Area:	managed mainly for the sustainable use of natural resources

Table 18. IUCN protected-area categories (cf. EEA 1995).

10,000 of the approximately 40,000 protected areas in Europe already comply with the criteria defined by the IUCN and are internationally recognised: of these, 65.5% were in category V in 1992 (WCMC), 14.9% in category IV, 11.8% in category II, 6.9% in category I and 0.9% in category III. The adoption of the categories in Europe facilitates comparisons of the protected areas in Europe as well as between those in Europe and the rest of the world. Worldwide the most numerous category is that of national parks, with 49.9%, followed by category IV, habitat- and species-management areas, with 30.3%. Protected land- and seascapes (category V) amount to 17.9% of the total worldwide; category I (nature reserves etc.) makes up 6.6%, category III (natural monuments etc.) 2.6%.

In *1985 the European Union* began a comprehensive programme for co-ordinating information on the environment: *CORINE* – Community-wide Co-ordination of Information on the Environment. One part of this programme was the compilation of a biotope registry for the EU *(CORINE biotopes project)*. The goal of this project is the "inventory of the most valuable ecological biotopes from a European standpoint in the member states of the EU" (cf. WASCHER 1993). This entails the establishment, cartographic and substantive description of internationally significant areas according to unified selection criteria. The CORINE Biotope Data Bank has inventoried and described 6,000 areas. At present the expansion of the CORINE Biotope Project to include the five Central European countries of the Phare programme is in the process of being worked out, so that data have begun to be compiled in Rumania, the Czech Republic, Slovakia, Poland, Hungary and Bulgaria.

Since May, 1992 the FFH Directive (Fauna-Flora-Habitat Directive/ FFHD) has gone into effect, bolstering *habitat protection* in the EU. Its main goal is "the preservation of biological diversity" and "(...) to maintain or restore a favourable state of preservation of natural habitats and wild animal and plant species of interest to the Community" (Art. 2/2). To this end, a "coherent European ecological network of special protected areas called *Natura 2000* will be established" (Art. 3/1). This system of protected areas comprises all the areas thus far certified by the EC Bird-Protection Directive and all the new ones which will be certified in future by the Bird-Protection Directive and the FFHD. The text of the guideline consists of three main sections: one on "habitat protection", one on "special species-protection regulations" and further regulations concerning execution, information and public relations.

The most urgent goal in the field of habitat protection is the co-ordination and improvement of the nature-protection regulations in the *Mediterranean countries*, particularly as regards unified species-protection regulations. For the protection of endemic species the main focus regionally is likewise concentrated on the endemites of the *Mediterranean and the Canary*

Islands, Azores and Madeira (cf. SSYMANK 1994, p. 396). The main focus of protection in Central Europe is clearly on the habitats named in Appendix I of the Directive. This ranges from *the marine and coastal* habitats (dunes, salt meadows, steep coasts) to standing and flowing bodies of water, heaths, numerous types of extensive meadowlands, marshes, natural rock formations down to forests and bushes. The habitats named here are largely in agreement with the "biotopes to be specially protected" stipulated by German nature-protection legislation (art. 20c BNatSchG – Federal Nature-protection Law).

The *implementation of the Directive* has varied widely from country to country. The main attention in the Mediterranean countries is on species protection, whereas in Central Europe the focus is on habitat protection. Whereas in *Germany* existing data of biotope charting are used to this end, some Mediterranean states, such as *Italy* and particularly *Spain*, have elected to conduct an active new on-site survey of the areas which meet FFHD criteria (cf. SSYMANK 1994). For the Mediterranean countries satellite data on landcover on a scale of 1:100,000, provided by the CORINE landcover project, are already available. In Spain a blanket mapping of vegetation on a scale of 1:50,000 has been begun which far exceeds the scope of German mapping projects. In Spain a clearly delineated survey of the areas of the FFHD will be available when the mapping project is completed. In *Italy*, too, further mapping projects are in progress, along with extensive plans for national parks to implement the Directive. In *Great Britain* there are already comprehensive expert assessments and surveys for nature conservation (SPSI), thanks to which some 900 areas have been legally protected since 1949, which will be a tremendous help particularly for the political and administrative implementation of the Directive. *In Spain not only was the system of protected-area types contained in Natura 2000 completely revised and adapted to current tendencies, but also a comprehensive nature-protection plan with regulatory as well as developmental considerations was worked out, with its main focus on protected-area planning* (cf. SCHMIDT 1995).

7 Threats to Coastal and Marine Ecosystems

In recent decades the nature of the coast has changed profoundly in many places. The natural processes and functions of the seas and coasts are being increasingly influenced and disturbed by anthopogenic intrusion. Coastal cities and harbours have spread out markedly. Agriculture, fishery, industries, coastal protection and tourism are often not subject to any restrictions and have an impact on the coastal system as a whole and are thus the source of a number of problems. *On no coast of the world is the environmental stress as great as in Europe* (EUCC 1992):

- large-scale pollution by oil spills from tankers or off-shore drilling rigs causing lasting damage to flora and fauna;
- in many places large areas have been transformed into harbours; transshipment operations and industries have polluted air and water in the greater environs; even today there are many industries which discharge their untreated sewage into the sea; thus, various chemical and often toxic substances are introduced to coastal waters and sediments (cf. op. cit.);
- high levels of freshwater removal cause salt water to encroach the near-coastal ground- and surface-water reserves;
- decimation of fish stock from overfishing, water pollution, destruction of spawning grounds etc.;
- increase of shore erosion, with resulting loss of land, as the sediment transport from land and shore is interrupted by buildings on the coast;
- increase of tourist activities, preferably in coastal segments which are (or were) in good ecological condition; on the Mediterranean coasts there are already 100 million tourists annually; according to estimates of the EUCC (1992) there could be as many as 250 million tourists by 2030;
- increasing activity by the coastal population in the fields of industry, agriculture, transportation, tourism and recreation has caused a shift in the general ecological balance and a concomitant reduction of adaptation and survival capacity of coastal ecosystems.

Human impacts on coastal and marine ecosystems generally do not have effects on specific biotic communities, but there is an increasing number of occurrences, such as, e.g., algae bloom or dramatic changes in the makeup of biotic communities, which are indicative of a widespread imbalance in European marine ecosystems (cf. EEA 1995).

7.1 Sources of Threat

Through the overlapping of the above-mentioned and other effects of human activity the system/process balance of coastal ecosystems has already been profoundly changed and disturbed. This poses a general threat to the existing resource and use potential of coastal zones. According to a study by RATHS, RIECKEN and SSYMANK (1995) on the threat to habitat types in Germany and their sources, a number of major sources of threat to marine and coastal biotopes can be named.

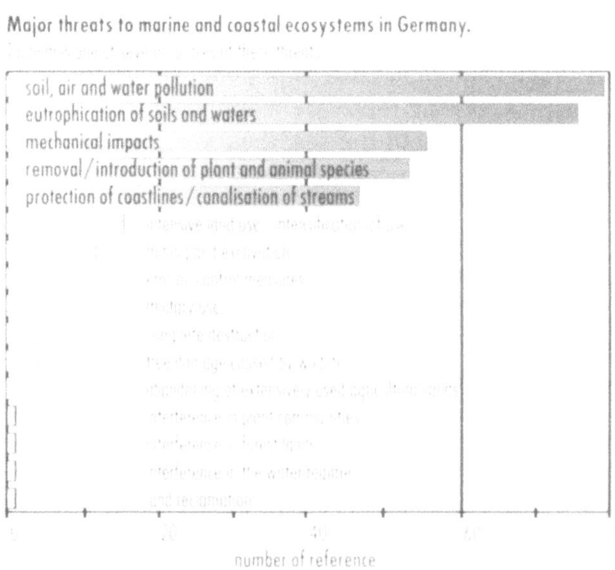

Fig. 12. Assessment of main causes of danger to marine and coastal biotopes. Source: RATHS et al. 1995, p. 209, altered.

The influences of human activity on coastal and marine ecosystems can be classified into three categories:

1. direct activity: degradation and destruction of habitats through land reclamation, building activity, bulldozing and damaging fishery techniques;
2. pollution: deterioration of the quality of coastal and marine habitats;
3. damage to biotic communities and loss of resources from non-sustainable removal practises (cf. EEA 1995).

The first source of threat for coastal and marine biotopes to be named is direct and indirect pollution from solid and liquid pollutants (e.g., heavy metals, toxic organic substances, petroleum etc.) which are transported there via the streams, but also directly (dumping, shipping, oil rigs etc.). Moreover, marine and coastal systems are subject to considerable impairment from the removal of marine animals and the mechanical impairments this involves, above all from ocean-floor trawling.

The threat to marine and coastal biotope types stem to a not inconsiderable extent from causes located in the interior, among them industry and commerce as well as agriculture, coastal protection, tourism and recreation (cf. RATHS et al. 1995).

7.2 Impairments of Coastal and Marine Ecosystems

Coastal zones and their valuable resources are regarded internationally, despite all attempts at protecting them thus far, as highly threatened and in danger of loss or destruction (cf. STERR 1993). The following list describes the major impairments of coastal and marine ecosystems currently occurring, according to the EEA (1995).

7.2.1 Water: Water Pollution and Algae Blooms

Water pollution coming from the land and diffuse pollution are the main causes for the general deterioration of water quality. It ranges from PCB pollution in the *Baltic* (WWF 1991) to oil and pesticides in the *Caspian Sea*. Even scarcely populated areas are impacted with water pollution spread there via the water or the food chain. In regions such as the *Mediterranean* there are major problems caused by organic pollutants (including sewage). The capacities of the water-treatment facilities are generally only insufficiently geared to the masses of tourists in the summer. Although the organic pollutants are quickly degraded, nutrient enrichment is one of the major causes for increasing frequency of algae blooms in recent years. Such algae blooms cause great problems, for example the bloom of *Chrysochromulina polyepsis*, which caused massive dying of salmon in the Norwegian salmon farms in 1988 (EEA 1995, p. 213).

7.2.2 Marine Fauna: Overfishing

Although it is difficult to establish it as a cause, it is probable that overfishing (e.g., in the *North Sea*) is responsible for great changes in the fish populations. In addition, overfishing leads to a lack of food for other animals, such as the striped dolphin *(Stenella coeruleoalba)* or the *Mediterranean* monk seal. Some fishery techniques have destructive impacts on populations which are not fished, such as fishery equipment being dragged behind the trawlers, which destroys coral biotic communities off *Norway.* Trawling nets are particularly dangerous to marine turtles and seals.

7.2.3 Salt Marshes: Draining and Overgrazing

Salt marshes are particularly threatened by drainage and reclamation projects to transform them into fields (polders) and pastureland. Approximately 50% of the invertebrate fauna are threatened by coastal-protection projects (EEA 1995, p. 213). Moreover, illegal dumping, landfills, aquaculture, tourism, hunting and changes in water level caused by civil-engineering projects damage and destroy salt marshes. Such factors are particularly damaging in the *Caspian* and *Black Seas*, where individual species are extremely dependent on certain salinity levels (BALKAS et al. 1990 in EEA 1995).

7.2.4 Estuaries and Deltas: Contamination

Estuaries, deltas and the biotopes associated with them are the *most endangered European ecosystems.* Estuaries have long been the preferred sites for harbours, moorings and land reclamation, or they are often used for garbage disposal. Industrial and urban waste borne by the rivers is transformed because of the changed salinity and energy conditions in the transition from fresh to salt water and deposited there as harmful compounds. Such waste water is often highly inorganic and becomes enriched or only decomposes at slow rates. Metals and chlorinated pesticide residues can reach dangerous concentrations in landlocked seas such as the Baltic (WWF 1991).

7.2.5 Decline of Sand Dunes Through Tourism

Sand dunes are found on all the coasts of Europe. The forms and structures of European dune landscapes vary widely from region to region and have a long history of development and human influences. *Fig. 13* shows the (approximate) total picture of sand dunes in Europe, as far as EUCC data were available.

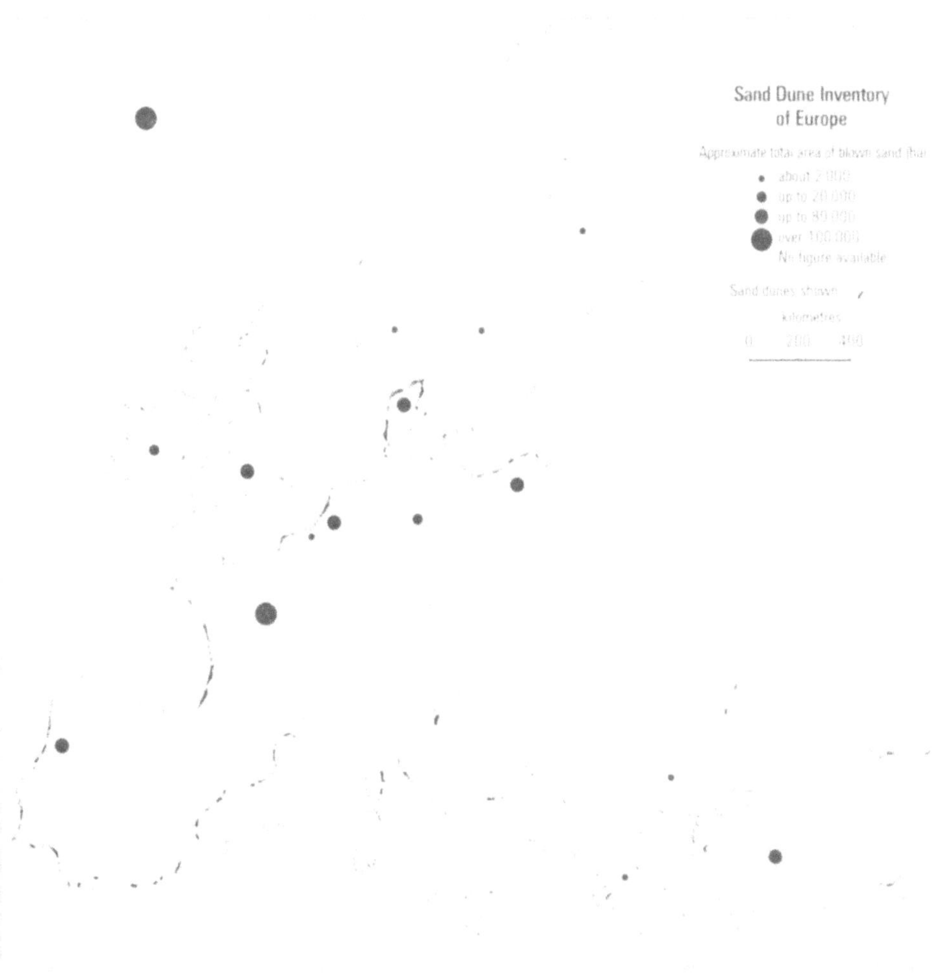

Fig. 13. Sand dunes in Europe. EUCC 1991, p. 5, altered.

A study conducted by the European Union for Coastal Conservation (EUCC) in 1992 resulted in alarming statistics on the change and loss of dune landscapes in Europe *(Table 19)*.

	Estimated area (ha) in 1900	Loss by 1990 (%)
Belgium	5,000	46
Denmark	80,000	35
France	250,000	40 - 50 (Atl); 75 (Med)
Great Britain	80,000	40
Greece	15,000 - 20,000	40 - 50
Ireland	20,000	40 - 60
Italy	35,000 - 45,000	80
Netherlands	45,000	32
Portugal	100,000	45 - 50
Spain	70,000	30 (Atl); 75 (Med)
Germany	10,000 - 12,000	15 - 20

Table 19. Sand-dune decimation in Europe from 1900 to 1990. Source: EEA 1995, p. 212, altered.

On the *Atlantic coast*, for example, there are still from 400,000 to 428,000 hectares of dune landscape today; 40% of the originally extant dune area (as of 1900) has been destroyed. Of this 40%, it is estimated that one third alone disappeared in the last three decades. Another 5% of the dune landscape will be destroyed in the next two decades, according to predictions of the EUCC, primarily due to leisure-time activities in the dunes, such as golf. A further loss of 10 to 15% is expected to take place due to massive activity in the leisure-time and tourism field in general and thus due to massive encroachment on dune landscapes.

Sandy beaches are major points of attraction for most of the 100 million tourists who frequent the *Mediterranean region* every year. It is precisely in the Mediterranean region that the enormous loss of an estimated 71% of the sand dunes extant in 1900 has taken place, caused by the construction of tourist facilities and the concomitant impact of tourism.

All in all, it may be said that direct activity in the coastal vicinity is the greatest factor in the shaping and endangerment of European coastal landscapes. Almost half of the sites shown in *Fig. 14* are threatened by one or more forms of encroachment. In the open waters pollution and resource extraction are the major problems. In near-coastal land areas it is construction activity, infrastructure steps and coastal-protection projects.

In some regions one or more major factors can be identified as the cause of certain damage and then tackled at the local level. Elsewhere coastal and marine ecosystems must be considered together and the entire spectrum of impacting human activity changed.

Coastal and marine ecosystems are among the most endangered landscapes in Europe. *Figure 14* shows 135 representative sites, which were evaluated according to their protective status, biotope management and predominant burdens. The selection was made based on international protected-area categories (such as the presence of biosphere reserves) as well as data on European landscapes in the data base of the WCMC and the CORINE projects. The result is that today in the *states bordering the Adriatic Sea* there is the greatest amassing of ecosystems which have thus far been exposed to all threats and stress because they had no protective status (denoted by red dots). Also, in the *Aegean Sea off the coast of Greece* and in the northern part of the *Caspian Sea* there are large areas with valuable but unprotected ecosystems which are subject to severe stress. The greatest amassing of burdened but protected marine and coastal ecosystems (denoted by orange dots) is on the coasts of *Denmark* and the *Geman coast of the North Sea*, as well as in parts of *France* and *northern Italy* on the Mediterranean coast. Protected, relatively unburdened ecosystems (denoted by green dots) are located primarily on the *German coast of the Baltic* and, again, on the Mediterranean coasts of *France* and *northern Italy*.

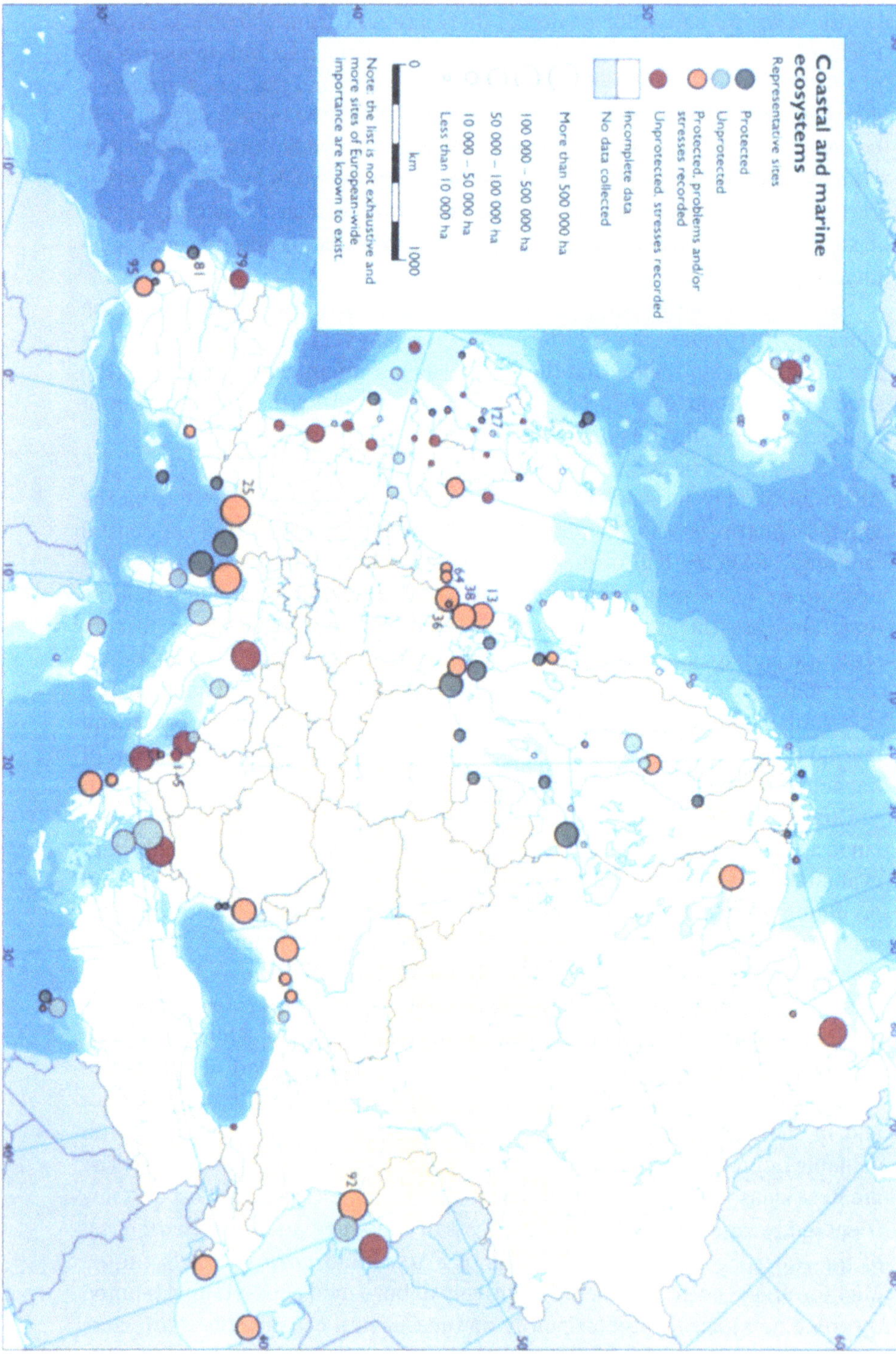

Fig. 14. Coastal and marine ecosystems: representative sites (EEA 1995, p. 214).

7.3 Overview of the Major Threats to the European Seas and Coasts

Not only do the abiotic and biotic characteristics of the seas and coasts of Europe vary region by region (as shown in chapter 5.3), the uses and threats to European coasts and seas on the part of the bordering states also vary widely. Below is a schematic listing of the major causes of threat to the European seas and coasts.

	Ecological state	Inorganic and organic pollution	Microbial contamination	Eutrophication	Other impairments
Mediterranean	severely impaired, particularly impacted the coasts of northern Spain, France, Italy and the Adriatic as well as coastal regions near large cities, e.g. Athens	yes, particularly from untreated household and industrial sewage during the tourist season (mainly caused by tourism!), from influx of river water transporting agricultural residues and from oil	yes*, severe contamination particularly on the coasts of the northwestern basin of the Mediterranean	yes*	building activity, overfishing
Black Sea and Sea of Azov	very severely impaired, as of now uninhabitable for most higher organisms. *Oxygen deficit between May and October in 60–80% of the surface area of the Sea of Azov and in the north western portion of the Black Sea* (area of 40,000 km^2) (cf. BRONFMANN 1993)	yes, from untreated household, industrial and agricultural sewage and oil, the latter poses a health risk for man	yes, from untreated sewage, health risk for man	yes, hypereutrophic nutrient influx from rivers and untreated sewage with resulting algae bloom, above average multiplying of jellyfish and habitat loss of benthic organisms due to lack of oxygen	dams and draining reduce sediment and freshwater via rivers, resulting in erosion in coastal areas and a change in salinity. Overfishing, changed in connection with eutrophication of food chains, very one-sided fish populations

Table 20.1. Main causes of threat to European seas and coasts. Source: *see Table 20.3.*
* many causes. n.d. = no data.

	Ecological state	Inorganic and organic pollution	Microbial contamination	Eutrophication	Other impairments
Caspian Sea	northern portion moderately polluted, southern portion very severely polluted	yes, from oil, untreated sewage and diffuse sources	n.d.	yes, oxygen deficit in deeper layers, changes food chains, oil extraction	water back-up (dams in the Bay of Kara Bogaz Gol). Decline of water level from reduced influx from the Volga
White Sea	high water quality, only locally impaired (in Onega and in the Gulf of Dvina), but coastal areas severely damaged	yes, by river water bearing industrial and agricultural sewage (especially synthetic organic material), in places oil as well	n.d.	n.d.	exploitation of natural eelgrass areas, reduction of seal population, rafting
Barents Sea	very slightly impaired	n.d.	n.d.	n.d.	from the sinking of radioactive waste and from global climate changes (ecosystems change from warming or lowering of temperature)
Norwegian Sea	high water and biological quality, slightly impaired only locally along shipping lanes and the coasts	yes (coast), from industrial wastes and (along shipping lanes) from garbage and oil	n.d.	yes (coast)	dangers from oil and gas extraction

Table 20.2. Main causes of threat to European seas and coasts. Source: *see Table 20.3.* * many causes. n.d. = no data.

	Ecological state	Inorganic and organic pollution	Microbial contamination	Eutrophication	Other impairments
Baltic Sea	very severely impaired (especially coasts and Kattegat) The state of the Baltic has worsened severely since 1950, pollutants contain increasing amounts of toxic substances	yes*	yes, leading to severe reduction of the recreational value of the Baltic etc.	yes (particularly on coasts) from high nitrate and phosphate concentrations and lack of movement in deeper water layers, leading, together with pollution, to fish dying, toxic algae blooms, species decline and changed species makeup, damage of seafloor vegetation	Overfishing "But the Baltic Sea is not only threatened by harmful substances and nutrients. The leisure time industry also demands its tribute and is laying claim to more and more room for yacht marinas and hotels" (BERLINER MORGENPOST, Sept 4, 1994, p 115)
North Sea	severely impaired	yes, especially heavy metals and synthetic organic substances	yes	yes, leading, coupled with pollution, to fish dying and algae bloom	potential threat from global climate changes and resulting sea level change much touristic activity on the coasts
North Atlantic	only very slightly impaired, coasts and Irish Sea more severely impaired	yes, along shipping lanes from garbage and oil		yes (along the coasts and in the Irish Sea), to an extent similar to the landlocked seas	overfishing (around Iceland and the Irish Sea)

Table 20.3. Main causes of threat to European seas and coasts. Source: calculations based on EEA data, 1995.
* many causes.
n.d. = no data.

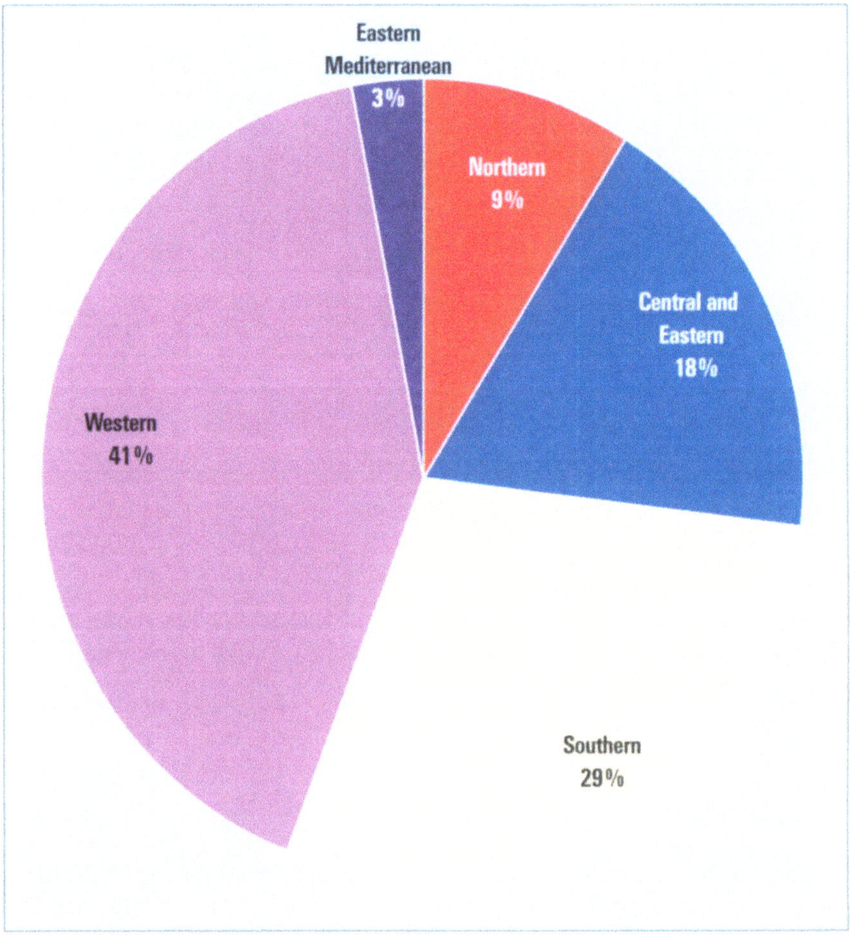

Fig. 15. Percentage of international tourist arrivals in the various regions of Europe. Source: EEA 1995, p. 490.

8 Coastal Tourism in Europe

According to the WTO (1993), the number of international arrivals in Europe has increased from 190 million (1980) to 288 million (1992). The average annual growth rate is 3.5%. But the masses of tourists are very unevenly distributed among the European countries, as shown in *Fig. 15*. Western Europe, with 41%, has the most tourist arrivals, followed by the southern regions (29%) and the countries of Central and Eastern Europe (18%).

Coastal areas are attractive tourist destinations throughout Europe. In the Mediterranean region alone, the most popular destination worldwide (according to the WTO), 35% of all international travellers are concentrated annually (EEA 1995). The attractive landscapes of the Mediterranean region, its cultural diversity, the pleasant climate and the numerous beaches have made it the leader in international tourism. 90% of the tourists streaming towards the Mediterranean are concentrated in the coastal areas of three countries: France, Spain and Italy. Between 1970 and 1990, tourism statistics in the Mediterranean region rose from 54 to 157 million travellers (EEA 1995). In some countries, Greece, for example, the number of arrivals increased six-fold during this period. In Turkey the number of tourists increased four-fold. Accordingly, in some tourist regions the capacities were vastly overextended. In the environs of Tarragona (Spain), for example, a tourism intensity of 4,250 tourists per square kilometre was found to exist in August, 1987 (CCE 1993).

8.1 Regions of High Tourism Intensity

At present the comparability of tourism-statistical data on the international and European planes is extremely limited. The statistical guiding principles and compilation methods vary widely from country to country. However, at the Bureau of Statistics of the European Union intensive efforts are now being made to co-ordinate tourism statistics (cf. Statistisches Bundesamt 1995). As no specific data for coastal regions are currently available, this means that the travel arrivals of foreign tourists available for a country as a whole must be adjusted for the coastal regions. This is also true of tourism intensity *(cf. Table 21 und Maps 8 and 9)*.

In the ranking of destination countries of foreign tourists, France is in first place among European countries (cf. Statistisches Bundesamt 1995, p. 179). With annual arrivals of nearly 60 million foreign travellers (1992), France is far ahead of Spain (with nearly 40 million) and Italy (approx. 26 million), followed by Great Britain, Germany, Greece and Portugal. The lowest tourist-arrival figures in Europe are in Finland, with 0.8 million registered arrivals, followed by Denmark, with not quite 1.5 million, Norway (2.3 million) and Belgium (3.2 million).

Tourism intensity in European countries produces a slightly different picture. Malta, with 3,171 tourists per square kilometre, has a tourism intensity far above average. Compared to other countries, the Netherlands display the highest tourism intensity (144/sq.km.), followed by France (110) and Belgium (106). The lowest tourism intensity is once again in Finland (2.3) and Norway (7.3). Concrete and comparable data to determine the specific tourism intensity in coastal areas do not exist.

Table 21. *(right)* Comparison of European tourism data. Source: calculations based on WTO data, summarised by Statistisches Bundesamt 1994.

[1] arrivals of foreign tourists in 1992, in 1,000s; Stat. Bundesamt 1994.

[2] tourists per km^2.

Europe/country	Tourism in absolute figures[1]	Tourism intensity[2]
Albania	**n.d.**	–
Belgium	3,220	**106**
Denmark (+ Faeroe Islands)	1,543	35.0
Germany	**15,147**	42.0
Estonia	**n.d.**	–
Finland	*790*	*2.3*
France	**59,590**	110
Gibraltar	**n.d.**	–
Greece	9,331	70.7
Ireland	3,666	52.1
Italy	**26,113**	86.6
Latvia	**n.d.**	–
Lithuania	**n.d.**	–
Malta	1,002	**3.171**
Netherlands	6,049	**144**
Norway	2,375	*7.3*
Poland	4,000	*12.7*
Portugal	8,884	96.1
Spain	**39,638**	78.5
Sweden	**n.d.**	–
Great Britain (+ Northern Ireland)	**18,535**	75.9
Yugoslavia (former)	**n.d.**	–
Turkey (Asia)	6,549	*8.4*

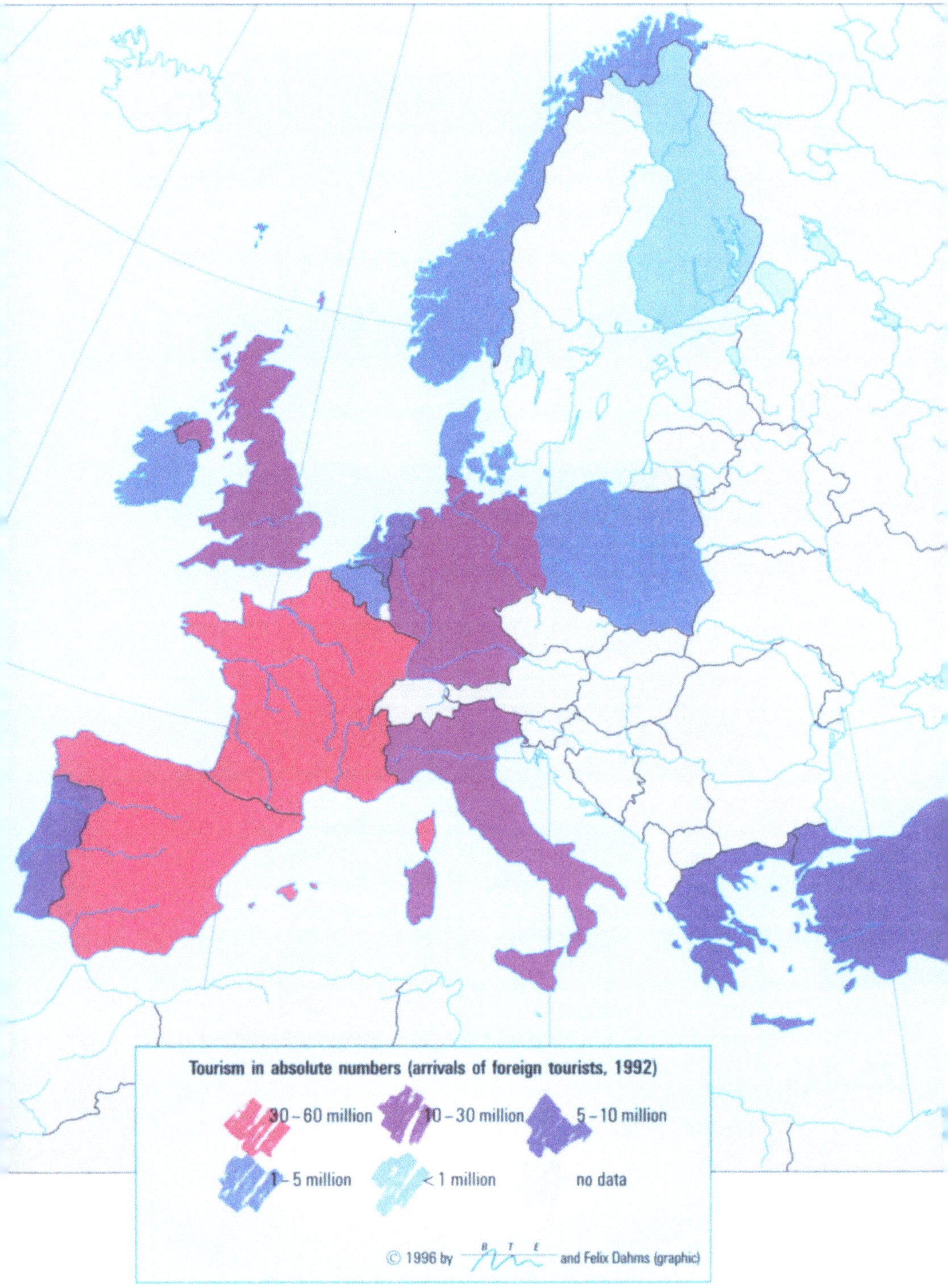

Map 8: Tourism in absolute numbers (arrivals of foreign tourists, 1992).

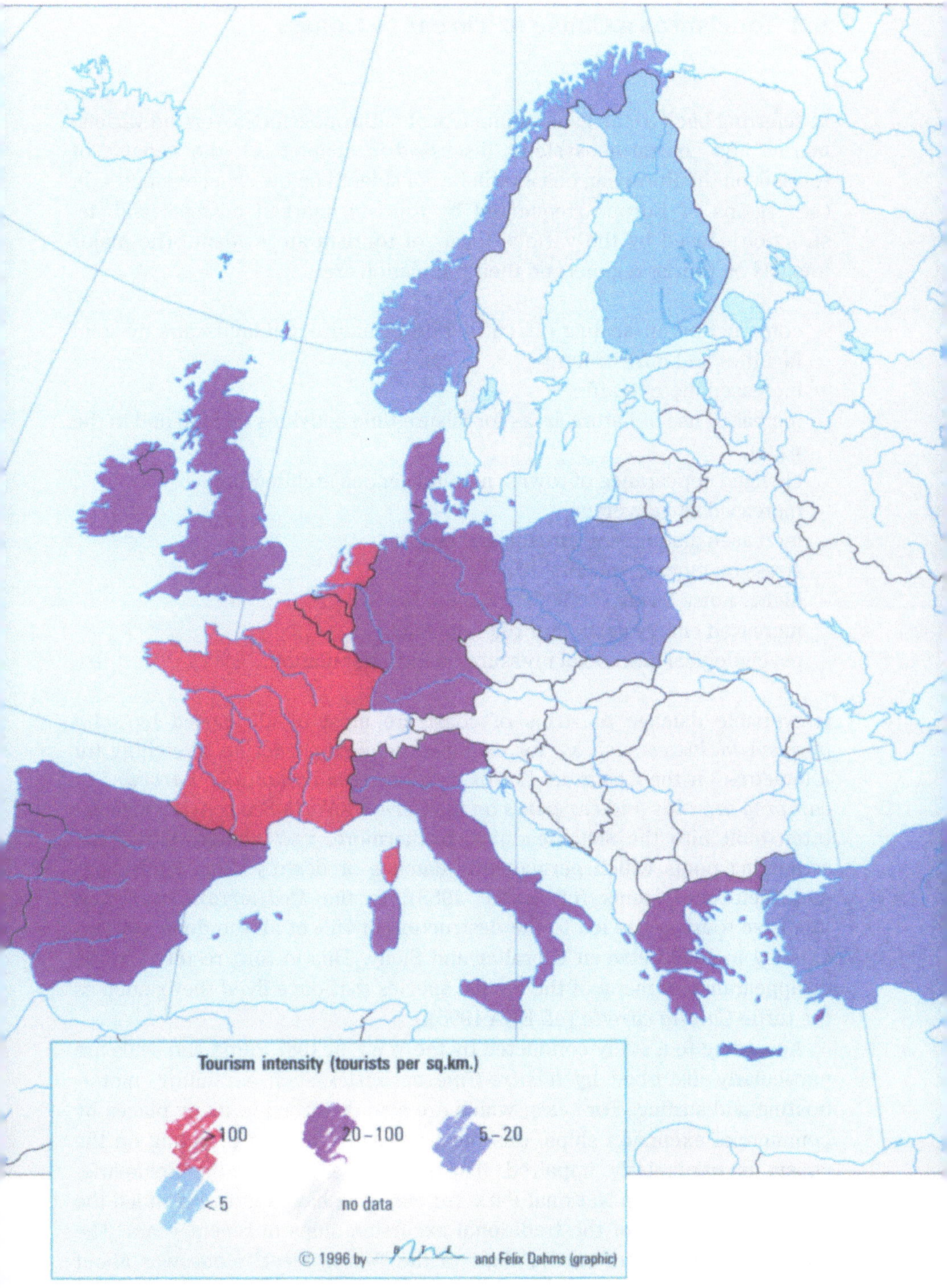

Map 9: Tourism intensity (tourists per sq. km.).

8.2 Tourism as a Cause of Threat to Coasts

In referring back to the general impacts of anthropogenic factors on various marine and coastal ecosystems discussed in chapter 3.1, the impacts of tourism on the European coasts will be considered below using examples. In the regions of Europe frequented by tourists, marked changes and destruction caused by the various forms of tourism are evident. The major impacts on tourist regions and their population are:

- construction on/sealing off/cutting off of nature and landscape by solid facilities and infrastructure,
- increased motor traffic,
- increased use of nature areas for leisure-time activities on land and in the water,
- changed appearance of towns, non-indigenous architecture,
- increased site coverage,
- increased amounts of garbage and sewage,
- stress on climate and air,
- higher noise levels,
- increased energy and water requirements,
- psychological and social pressure on local population.

Irreversible damage to strips of coast are most of all caused by *solid tourism facilities* such as, for example, *hotel construction* stretching for kilometres on the Costa del Sol or *small-boat marinas and harbours or camping grounds and car parks* on the *German North Sea coast*, which are often built into the salt meadows; furthermore, vast holiday parks and swimming pools which permanently damage or destroy dune landscapes and their ecosystems (cf. LAMP 1988). In the *Mediterranean region* intensive tourism has led to the destruction of 75% of all the dune systems existing in 1900 between Gibraltar and Sicily. This in turn resulted in the disappearance of many of the animal species that once lived there, such as the turtle *Caretta caretta* (cf. EEA 1995).

According to a study conducted by the WWF in 1987, birds and seals are particularly disturbed by *leisure-time activities* such as sailing, motor-boating and surfing. For seals, which are also disturbed in many places by commercial excursion ships, this means that the rearing of young on the coasts is particularly impaired. The ships operating in the Schleswig-Holstein Wadden Sea National Park, for example, have vastly expanded the passenger capacity of the traditional excursion ships in recent years. The number of passengers on the ships of the "White Fleet" alone was about 700,000 in 1991 and may have exceeded the one-million mark by now, according to ANDRESEN (1993). As to birds, touristic activities are very disturbing to nesting (e.g., on salt meadows and beaches consisting of crushed seashells in early summer), feeding (throughout the year), molting (August) and resting at high tide (cf. LAMP 1988).

A second group of *touristic activities* leading to environmental impairments consists of walking, hiking, horseback riding and bicycling in ecologically sensitive coastal areas. Plover nesting, for example, has been greatly reduced in many countries in recent decades (e.g., *Sweden, Hungary, Wadden Sea*) or disappeared completely (e.g., *Norway, Britain* and the *Baltic*). Increasing tourism and the destruction of habitats (by building over entire strips of the coast, trampling etc.) are considered to be the major causes of this decline (cf. STOCK & SCHULZ 1991). The largest plover colony in Northwest Europe is located on the sand dunes on the foreshore of St. Peter-Böhl on the Eiderstedt Peninsula. There up to 200 pairs – half of the entire plover brooding stock in Northwestern Europe – come to nest. But the white beaches also attract millions of holiday tourists every year. The demands of nature conservation and tourism, the most important sector of the economy in the region, converge head-on on the coastline of St. Peter-Ording (North Sea coast, Germany). Not even 10% of its total area is in the most strictly protected Zone 1 of the Schleswig-Holstein Wadden Sea National Park. Quite to the contrary: a car park is on the beach of the National Park in St. Peter-Ording, along with the traditional canopied wicker beach chairs and a bathing beach. But nesting plovers need an undisturbed buffer zone of at least 100 m around their nests, as studies on tourist activities in and around nesting areas in the National Park have clearly shown. Prohibiting access to small areas does not appear to be a suitable means of securing beach habitats (cf. ÖKOSYSTEMFORSCHUNG SCHLESWIG-HOLSTEINISCHES WATTENMEER 1994).

A further important group of environmental threat caused by tourism is the *excessive strain on utilities*, which is particularly significant on islands, as the supply of drinking water, for example, is often taken from the island's own reserves. The increasing water consumption in the summer exceeds capacity in many places (LAMP 1988). By now there are serious bottlenecks in the water supply on *Majorca*, for example. Thus, in the summer of 1995, drinking water had to be brought to Majorca on tankers. The Majorcans who pump their water also complain of an increasing salinity of the ground water, particularly in the two regions most strongly frequented by tourists, Palma and Calvia (DER SPIEGEL, no. 19/ 1995, p. 170 ff.). In *Turkey* high water consumption by tourists is forcing many communities to divert the water of remote mountain regions, intruding into ecosystems which are thereby indirectly impacted by tourism. The waste water is generally discharged untreated into the sea or into outfalls (OLDENBURG 1988).

In most coastal towns and on some islands, stress stemming from *traffic and the noise associated with it* and additional *areas devoted to traffic* are also major factors. By steps and initiatives such as limiting traffic and giving pedestrians priority, speed limits, bypass roads, car parks outside of town, inexpensive bus service and gratis one-day tickets and the promotion of bicycle traffic, attempts at reducing such stress are being made in many places. The banning of cars such as on Wangerooge or Juist in Germany have been the exception up to now (cf. ECOLETTER 1995, p. 19).

The impairment from *air transportation* is also high in some places; thus, on the *East Frisian Islands* of Borkum, Juist and Wangerooge between 15,000 and 35,000 flight movements were counted in 1985. In Schleswig-Holstein this figure is only exceeded in Sylt (LAMP 1988).

8.3 Regions of High Conflict Potential

When high tourist intensity (tourist pressure) on the coasts converges with high biodiversity, there is high conflict potential. When tourist pressure and biodiversity on the coasts is low, the converse is true: the conflict potential is low. The result of our calculations, based on data of the IUCN and the WCMC, but limited to the European region alone, is shown in *Table 22:* in the countries of Northern Europe tourist pressure on individual countries, in keeping with the figures on arrivals and tourist intensity, tends to be low (the only exception being Great Britain, which has a high ranking); in Western Europe it is predominantly in the medium range (exception: the Netherlands, also with high ranking); in the Mediterranean countries, in contrast, it is high throughout. An increase of the streams of the tourists from north to south is clearly discernible.

As it is precisely the Mediterranean countries which have a high biodiversity, i.e., which are considered particularly sensitive to encroachment, the Mediterranean region is also that European region with the greatest conflict potential *(cf. Table 22/Map 10)*. From this fact an urgent need for action in the Mediterranean countries may be inferred. However, a medium or low conflict potential should not be seen as suggesting that in these European countries there is no need for regulation nor tourist pressure impacting nature and the environment of coastal areas. Numerous studies from these countries prove that here, too, there is a need for action, as will be demonstrated using the exemplary cases (Section C) to illustrate each "conflict class".

Europe/country	Biodiversity (total[1])	Tourism "pressure" on European scale[2]	Conflict potential on European scale[3]
Albania	medium	n.d.	–
Belgiun	medium	medium	medium
Denmark (+ Faeroe Islands)	medium	low	low
Germany	medium	medium	medium
Estonia	medium	n.d.	–
Finland	low	low	low
France	high	high	high
Gibraltar	low	n.d.	–
Greece	high	high	high
Ireland	low	medium	low
Italy	high	high	high
Latvia	low	n.d.	–
Lithuania	low	n.d.	–
Malta	high	very high	high
Netherlands	medium	high	medium
Norway	medium	low	low
Poland	medium	medium	medium
Portugal	medium	high	medium
Spain	high	high	high
Sweden	medium	n.d.	–
Great Britain	medium	high	medium
Yugoslavia (former)	medium	n.d.	–
Turkey (Asia)	high	medium	high

Table 22. Tourism and biodiversity in European coastal regions: current potential conflict countries (ranking based on WCMC data).

[1] *see Table 15*

[2] arrivals + tourism intensity *see Table 21*

[3] biodiversity + tourism pressure

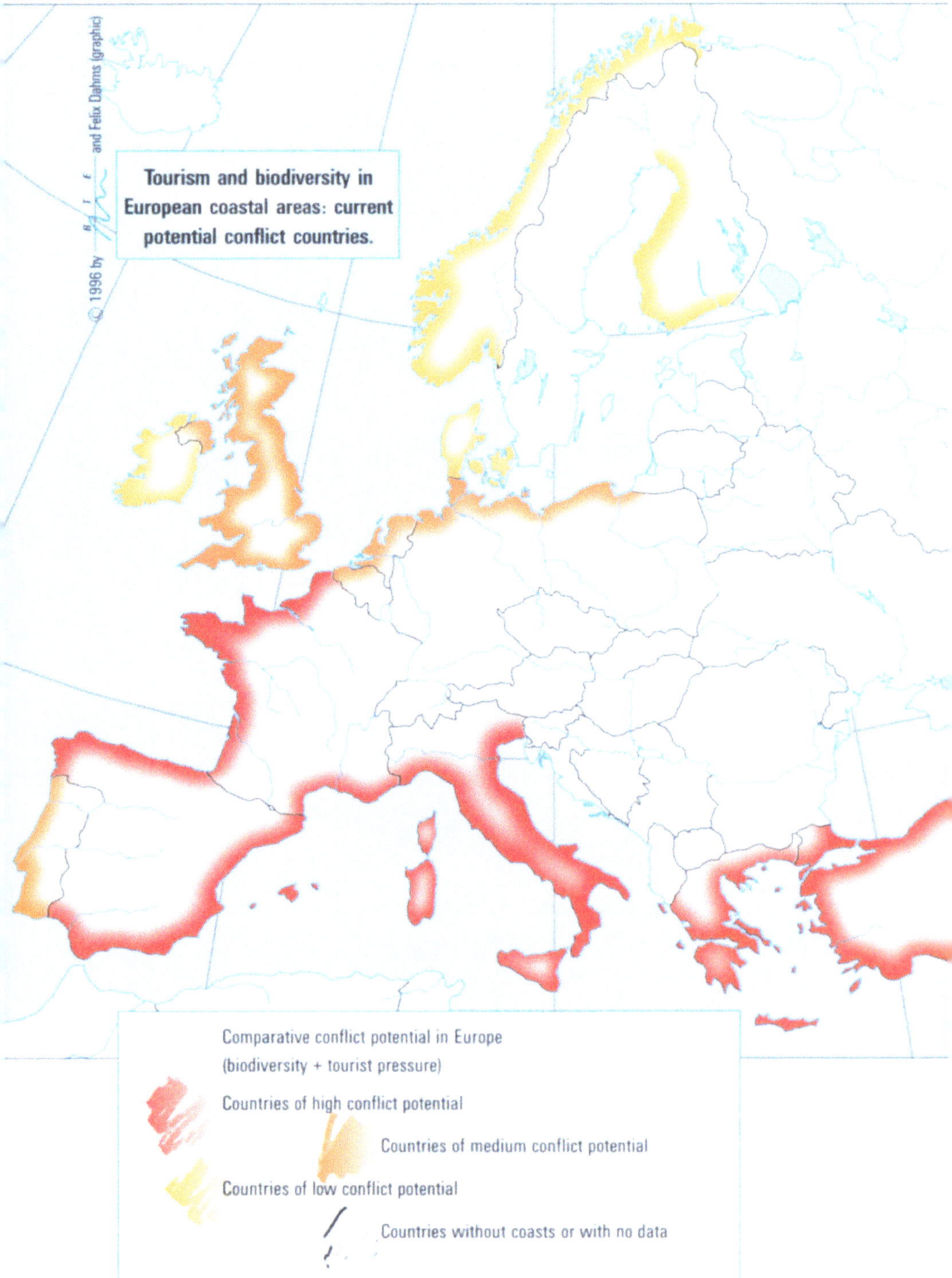

Map 10: Tourism and biodiversity in European coastal areas: current potential conflict countries.

8.4 Coastal Tourism and Its Impact as Exemplified by the Mediterranean

As the stress on the coasts caused by tourism is particularly severe in the Mediterranean region and demands quick solutions, a study was conducted in 1984 for the *Mediterranean region* (as part of the *Mediterranean Action Plan, MAP*) in which the impact of tourism on the environment was examined and described: *THE BLUE PLAN*. The focus of the study, which was supported by the UNEP, was concentrated on (cf. chapter 10):

1. *resource consumption* (i.e., land and water consumption) through tourism
2. *stress from emissions* and *waste*
3. *physical* and *socio-cultural stress.*

8.4.1 Site Coverage

The influence of tourism on site coverage is examined in the study using the indicators number of beds and actual overnight stays. It is assumed that every hotel bed takes up 40 sq.m. of area (immediate hotel area) plus further accommodation area of 70 sq.m. (including garden, parking etc.). *Table 23* shows the area taken up by the overnight capacities of the year 1984 (unfortunately, there are no more recent data; all studies of the Blue Plan are based on the data of 1984).

From this it is evident that in 1984, 90% of all the capacities were in the northwestern Mediterranean region. With the exception of Greece *(and Turkey: authors' remark)*, where there is likewise high tourist pressure today, this spatial polarisation in the Mediterranean region probably still holds true today. Thus, in the northwestern Mediterranean alone, 2,200 sq. km. were taken up by tourist accommodations. In addition, the area taken up by required touristic infrastructure and the usual concomitant urbanisation processes is doubled to 4,400 sq. km. Fifty per cent of the hotels included in the study are in the immediate coastal area of the Mediterranean. Taking up an area of 2,200 sq. km. is equivalent to a densely populated coast 1 km deep and 2,200 km long or a densely populated one 500 m deep and 4,400 km long. This illustrates the magnitude of the area taken up on a strip of coast.

	No. of hotel beds (1,000s)	Site coverage (1,000 m^2)	No. of other beds (1,000s)	Site coverage (1,000 m^2)
Spain	840	33,600	8,923	624,610
France	1,590	63,600	10,894	762,580
Italy	1,598	63,920	5,824	409,780
Malta	14	560	41	2,870
Yugoslavia	319	12,760	1,127	78,890
Greece	323	12,920	324	22,680
Turkey	68	2,720	182	12,740
Cyprus	27	1,080	137	9,590
Syria	23	920	71	4,970
Israel	65	2,600	162	11,340
Egypt	48	1,920	218	15,260
Libya	9	360	24	1,680
Tunisia	72	2,880	77	5,390
Algeria	27	1,080	37	2,590
Marocco	59	2,360	152	10,640
total	**5,082**	**203,280**	**28,223**	**1,975,610**

Note. The different types of tourist accommodations (hotel and other, including related areas) occupy an average surface of 25–100 m^2 per bed.

rented, self catering accommodation	50 m^2	per bed
hotels	30 m^2	per bed
youth hostels	30 m^2	per bed
holiday villages	100 m^2	per bed
camping and caravan sites	50 m^2	per place
car parks	20 m^2	per place

The average reached is 40 m^2 per bed for hotel accommodation and 70 m^2 per bed for other accommodation.

Table 23. Capacities and site coverage by hotels and other types of accommodations, 1984. Source: GRENON/BATISSE 1989, p. 155, altered.

8.4.2 Water Consumption

For water consumption, too, the indicators used are the number of beds and actual overnight stays. For the year 1984, an average water consumption by foreign tourists was assumed to be 250 litres per person per day, and for domestic tourists 150 l per person per day. The study assumes that tourists in hotels of the luxury class consume up to 600 l per person per day, whereas campers use hardly any (studies by SCHERB 1975 and the ADAC 1992, however, assume for camping sites with good sanitary facilities a per-capita water consumption of 145 l/person/day, cf. BTE 1995a, p. 37). For the BLUE PLAN this results in the average value given above.

The increased water removal leads to a sinking of the groundwater table, such as, for example, in the Hammamet region in Tunesia, where it resulted that the land dried out, was no longer arable and finally abandoned. In coastal regions this problem is exacerbated by salt water seeping in. Acute problems exist on the Balearic Islands and in some coastal regions of Spain.

8.4.3 Garbage, Sewage and Emissions

The *volume of garbage* was estimated on the basis of overnight-stay statistics; the *volume of sewage* on the basis of water consumption: per tourist a volume of garbage of 0.9 kg/day was arrived at and 60% of daily water consumption was assumed to be sewage in need of disposal, which, in the case of foreign tourists, amounts to about 150 l/person/day.

The influence of tourism on noise, atmospheric pollution from exhaust fumes and pollution of coastal waters cannot be quantified, as there were no comparable statistical data for the Mediterranean countries *(and there still are none: authors' note)*. Moreover, emissions overlap with waste products from other impacting factors.

Air pollution is concentrated primarily along the main overland roads. It should therefore be assumed that air pollution is that much higher in regions which tourists travel to in their cars. According to GRENON/BATISSE (1991), the most heavily impacted destinations are in Spain, France, Italy and the former Yugoslavia. In 1985 Yugoslavia, with 86%, had the highest number of trips with automobiles. In addition, traffic was concentrated on the coasts. In Syria, Turkey and Morocco, too, travel by car was also an important factor.

The tourists' *waste water* is discharged with the regular household sewage into sewage-treatment facilities. Polluted coastal water from discharge of untreated or inadequately treated sewage has already resulted in health hazards in many places, so that meanwhile data on water quality have been published by various bodies.

8.4.4 Excursus: Quality of Bathing on Europe's Beaches

Europe's bathing beaches are regularly tested for bacteria and viruses. According to the *EU norm* (EU Bathing-water Directive of 1975), a coastal segment is "good" if less than 100 certain coli bacteria are contained in 100 ml of water. If this level is exceeded, the water is only of medium bathing quality. A beach is classified as "bad" as soon as the limit of 2,000 bacteria per 100 ml is reached (KRAFT 1994). The environmental experts of the EU Commission classified some 90% of the European coasts as suitable for bathing (KÖRBER 1995). In the 12th Bathing Report of the Commission, based on water samples of the 1994 season, 17,000 bathing sites in the then 12 EU member states were tested. Data for the new members Finland, Austria and Sweden were not yet available. According to these tests, the cleanest water by far was in *Ireland*, where all samples met the EU minimum standard. *Spain* and *Greece* also had relatively clean beaches; about 95% of the strips of coast were classified as fit for bathing. In *Italy* and *Portugal* the share was about 85%, in *France* about 90%. Bad results were shown above all by the states bordering on the North Sea: in the Netherlands only 63% of the water samples were of sufficient quality; in Germany 80% of the strips of coast were classified as fit for bathing.

The largest German automobile club *ADAC* has set up a service which summarises the EU test results and constantly brings them up to date. In 1995, the ADAC itself tested an additional 3,400 European beaches on coasts and inland bodies of water, testing the water quality three times at each site. However, the test only includes beaches which are frequently visited by German tourists and can be reached by car: the *North Sea and Baltic coasts of Germany, the Dutch coastal waters, the French coast with Corsica, in Spain the coasts of Valencia and Catalonia, in Italy Liguria, Tuscany, Emilia-Romagna, Venetia, Friuli/Julian-Venetia and the Slovenian and Croatian coasts*. Only about 90 of the 3,400 beaches tested were classified as hygienically contaminated by the marine biologists (cf. ECOLETTER 9/95, p. 17). The quality of the bathing water has in particular clearly improved in the Mediterranean region, according to marine biologists. Reasons for this are the construction of new sewage-treatment facilities and the systematic expansion of the sewer network.

Together with its partners, *ECOTRANS* also tested the Germans' favourite European beaches. Here detailed data were gathered in 16 countries on water quality, beach service, environmental protection in the holiday setting etc. The data were based solely on voluntary information of the responsible authorities. An overview of the final results was presented in the *ECOLETTER* (9/95, appendix). The results showed that in particular *Turkey*, *Greece* and *Portugal* were "lagging behind" the environmental activities of the Northern European bathing destinations and those of Italy.

Apart from numerous buildings in coastal areas that should never have been built, there are still considerable weaknesses in sewage treatment and garbage disposal; also, information from there is generally insufficient. Half of the bathing sites (of 100 bathing destinations in the survey) have fewer than 10,000 inhabitants; but some 50% of these towns have over 1 million overnight stays annually. Even though things have improved in recent years regarding sewage disposal in the Mediterranean, eight of the towns under study had no access to a sewage-treatment plant and nine only to one with mechanical sewage treatment. Only in two thirds of the European bathing destinations is garbage separately collected, primarily glass, paper and metal. 16 of the sites, most of them in the Northern European region, sort 100% of their garbage (ECOLETTER 1995, no. 5/6).

8.4.5 The Stress on Man, the Environment and Landscape

In assessing the burden on man, the environment and landscape, the BLUE PLAN merely infers from tourism intensity a similar pressure exerting stress on the local populace and nature. *Table 22* provides an estimate of tourist pressure on the coasts of the Mediterranean for the year 1984. According to it, greatest tourist pressure was concentrated on the coasts of a few Mediterranean countries, above all Spain, France and the former Yugoslavia and the island of Malta, which closely approximates today's figures (cf. 8.1).

In these calculations the actual and often very great numbers of tourists visiting certain places which are considered particular favourites to sightseers, such as the historic or archeological destinations Venice or Luxor, or the regions travelled to because of their natural beauty, including many regions which have been placed under special protection, do not figure in. Thus, the Federation of Nature and National Parks of Europe (FNNPE) found in 1993 that there is not a single example of a truly environmentally and socially compatible tourism plan in the 40 protected areas under study. On the contrary, the masses of tourists have largely uncontrolled effects on the protected areas and thus constitute an extreme disturbance (cf. FNNPE 1993).

Section C. Exemplary Cases of Conflicting Use and Solution Strategies in European Coastal Areas

Potential conflicts of interest between tourism and biodiversity in Europe and the concrete impacts on the European coasts brought about by them were already described in chapter 8. The present-day conflicting factors on the coasts will herein be described using exemplary cases taken from regions of various European countries. The cases are selected such that countries from all three conflict potentials (low, medium, high – cf. chapter 8.3) are represented. Also, the selection is geared to reflect as great a range of current impairments from tourism on European coasts as possible.

The example of the Mediterranean region already cited (chapter 8.4) clearly shows that great masses of tourists, a lack of planning in the past and in part inadequate waste-disposal infrastructure have already resulted in a wide range of environmental stresses and destruction.

Tourism is a major economic factor in many regions, as the following descriptions of tourism development in the individual examples once again illustrate. And *it is precisely the touristic sector of the economy which lives off its capital, the intact nature and environment of the various tourism regions and a hospitable, friendly population* (DRV Umweltempfehlungen a, p. 2). The need to act is thus not only evident in the view of nature- and environmental-protection organisations and initiatives, the institutions of the "tourism sector of economy" are also challenged to act, as has been realised by a great many bodies. Strategies, steps and solution plans can be developed and implemented at varying levels: internationally (e.g., the UN, OECD, WTO, EU), nationally (e.g., ministries, marketing organisations, tourism associations, hotel associations etc.), at state level (e.g., state tourism associations, promotion boards), regions (e.g., regional tourism associations, regional hotel associations), communities (clubs, initiatives, tourist information) etc. (cf. op. cit.).

The following examples from various countries provide a first impression of the diversity of national and regional solution strategies which will then be complemented by national solution strategies for developing sustainable tourism in European coastal regions (chapter 10).

9 Exemplary Cases

The following overview of the exemplary cases illustrates the major factors involved:

Example Country/region	Conflict potential	Sources of risk and impairments (described in example)
9.1		
France/Côte d'Azur	high	vacation homes, conference tourism, leisure time activities, building over of dunes, coastal protection projects. ■ severe coastal erosion, ■ destruction of dunes
9.2		
Spain/Costa del Sol	high	construction of tourist facilities, in particular large hotel complexes, club areas, golf courses, marinas, construction projects on coast, ■ landscape depletion, ■ overdevelopment, ■ resource depletion, ■ destruction of dunes, ■ coastal erosion, ■ pollution

	Example Country / region	Conflict potential	Sources of risk and impairments
9.3	**Turkey/southern coast: Köycegiz (A), South Antalya (B)**	**high**	**construction of infrastructure facilities and hotel complexes and holiday villages, nature tourism (sea turtles as tourist attraction), boat tourism** ▪ damage to sea turtle population, ▪ resource depletion, ▪ problems with garbage removal, ▪ problems with domestic urban development
9.4	Germany Baltic Sea: Rugen	medium	expansion of tourist and leisure time infrastructure, island tourism, camping, much automotive traffic, unregulated parking ▪ overdevelopment, ▪ destruction of coast protective forests, ▪ impairment of nature reserves, ▪ visual impairment of landscape
9.5	Germany North Sea Schleswig-Holstein Wadden Sea	medium	construction of hotels, apartment houses, roads, harbours/marinas, bathing beaches spoiled by seawash, leisure-time activities, camping facilities ▪ additional emissions (sewage, garbage), ▪ damage to dunes, ▪ treading on vegetation, ▪ disruption of habitat of birds, seals and fish.

	Example Country / region	Conflict potential	Sources of risk and impairments
9.6			
	Ireland/northwest coast	low	intensive building activity, leisure time activity, golf in the coastal dunes, vacation homes, house trailers, marinas in bays and estuaries, beach car parks ▪ destabilisation and destruction of dunes, ▪ loss of marshes and lagoons, ▪ removal of coastal sand

9.1 Example: French Mediterranean Coast – Côte d'Azur

9.1.1 Touristic Development

The Côte d'Azur is France's most highly frequented tourist region. Over 8 million people come to the French Mediterranean coast every year. Monaco alone attracts over 3 million tourists every summer. Through the former fishing village of St.-Tropez, which has only 6,400 inhabitants itself, up to 100,000 daytime tourists pass during the main season. The "Blue Coast", as the approx. 300 km long strip of coast between Menton, on the Italian border, and Cassis, at the gates of Marseilles, is called, is one of the world's most expensive holiday areas. In the tourism sector of this region, three times as much money was earned in 1993 as in agriculture. 25 to 30% of the local workforce earn their livelihood in tourism. The popularity of the region is not only evident from annual revenues of approx. 25 thousand million FF, but also in the some 300,000 *holiday dwellings* located in this region. In addition, since the mid-80's a marked *conference tourism* has developed in some cities of the Côte d'Azur (Nice, Monte Carlo, Cannes), and it, too, has become a significant economic factor in the region. The congress and conference season, which begins in September, brings in a lot of money for the hotels and restaurants located there: statistics show that conference participants spend five times as much per day as holiday tourists (SANDBERG 1994).

The extremely rapid development of tourism on the Côte d'Azur is reflected in the degree of urbanisation of the coastal landscape. The development of urbanisation is illustrated in *Fig. 16*, using the region between Ollioules und Hyères as an example, which is one of the most popular tourist destinations on the Côte d'Azur. The illustrations show the development in 1960, 1977 and 1990 (cf. CONSERVATOIRE DE L'ESPACE LITTORAL 1995).

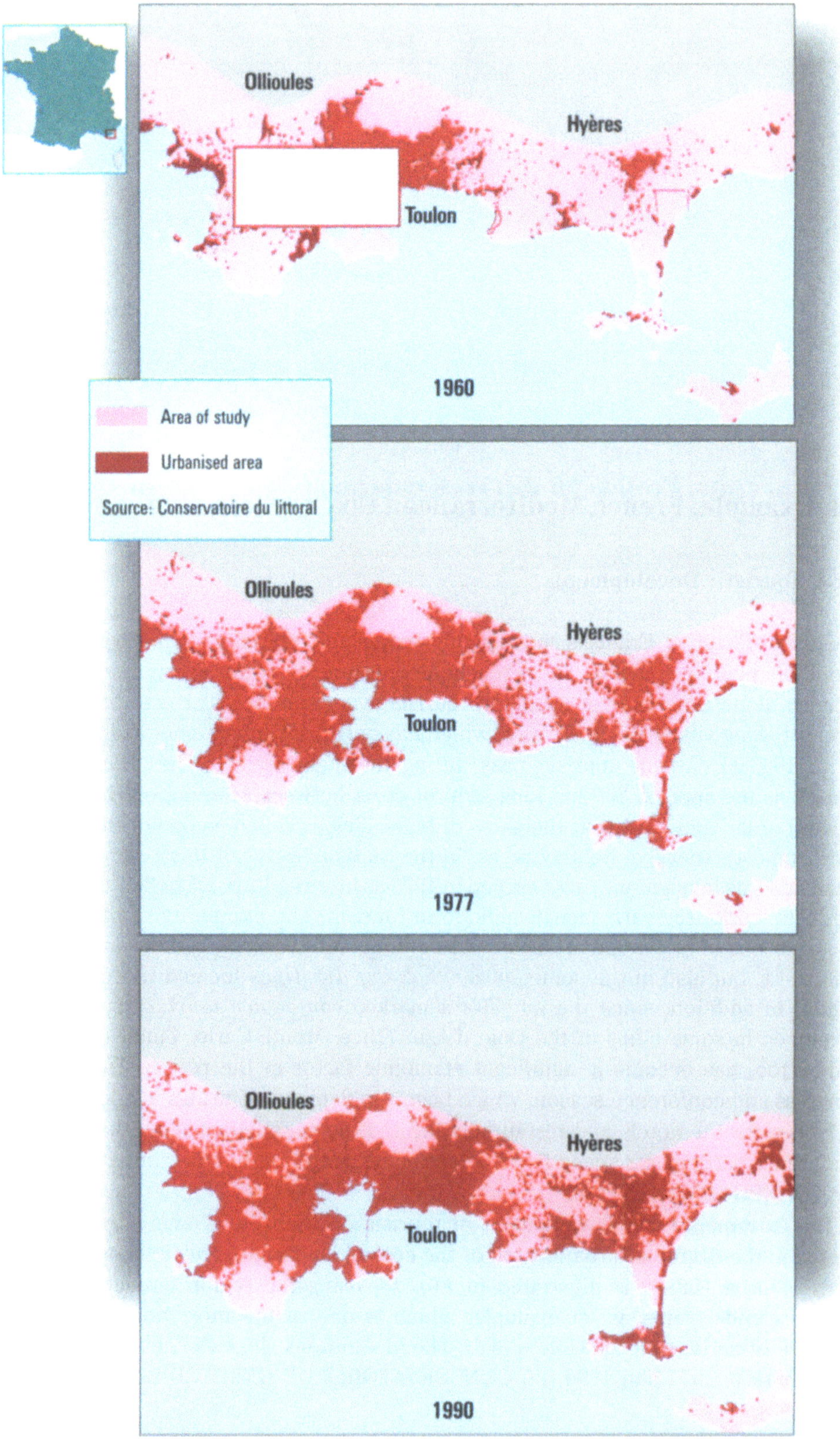
Ollioules
Hyères
Toulon
1960
Area of study
Urbanised area
Source: Conservatoire du littoral
Ollioules
Hyères
Toulon
1977
Ollioules
Hyères
Toulon
1990

9.1.2 Particular Natural Features

France has one of the most extensive dune landscapes in all of Europe. The largest areas are located on the western Atlantic coast and on the Côte d'Azur (cf. EUCC Sand Dune Inventory 1991, p. 43).

The coastal vegetation of the Atlantic may be compared to that in Northwestern Europe, which is characterised by a clear dune zoning and a rather "closed" character (as opposed to wandering dunes). The further south the dunes are located, the more elements of the typical southern, very species-diverse Mediterranean vegetation are evident (op. cit.).

Figure 17 provides an illustration of the major nature areas in France. The type I zones denote areas with extensive ecosystems worthy of particular protection, with a diversity of species and habitats. The type II zones are areas containing "nature ensembles" worthy of protection and isolated natural monuments. Especially in the northern portion of the Côte d'Azur, between Monaco and Frejus, there is a concentration of *type I zones;* it is the largest contiguous area with particularly protection-worthy ecosystems on the coast (and the second-largest in France as a whole). The very mild and temperate climate the year round on the Côte d'Azur is the reason that there are so many floral and faunal species. *Type II zones* also occur along the entire coastal strip of the Côte d'Azur.

Fig. 16. *(left)* Urbanisation process between Ollioules and Hyères on the Côte d'Azur. Source: Conservatoire du Littoral 1995, map 68; altered.

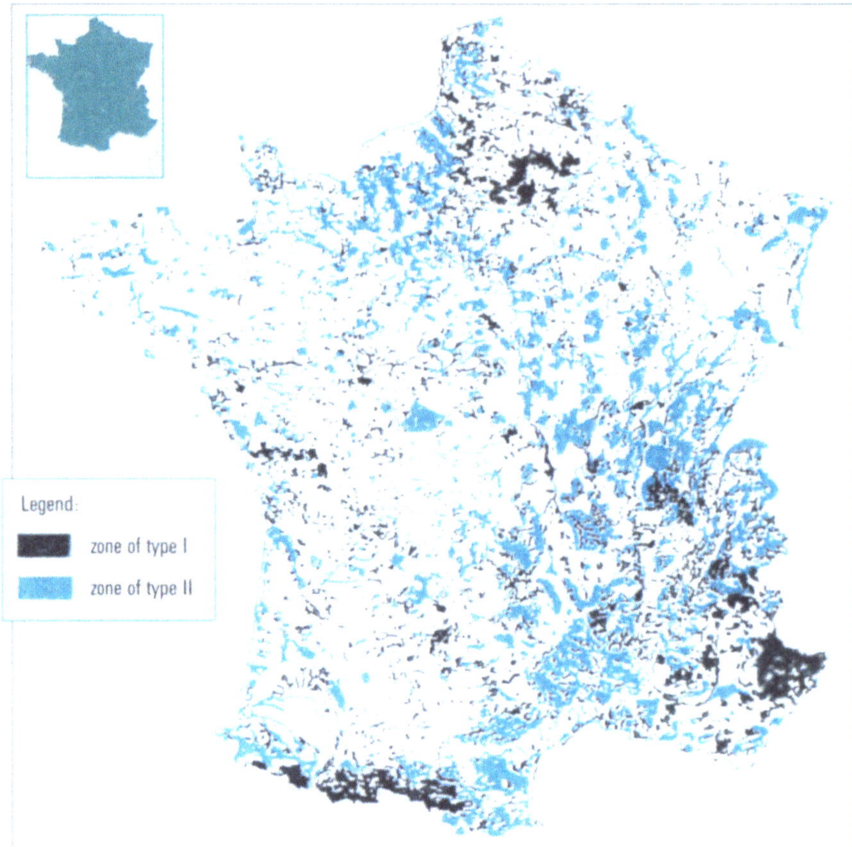

Fig. 17. Type ranking of French nature areas of ecological, faunal and floral interest. Source: ifen 1995, p. 372, altered.

The typically Mediterranean vegetation includes eucalyptus trees, cypresses and cedars, umbrella pines, plane and palm trees, of which there are over 60 species on the Côte d'Azur alone. Also, orange, lemon and almond trees are cultivated here. Many of these species were only introduced here in the last century, however.

The French coast between Menton and Cassis features geologically very differently formed coastal segments. Whereas between Cassis and Toulon it is characterised primarily by light-coloured limestone rock sometimes dropping steeply to the sea, sometimes interrupted by bays, the tree-covered Maures Massif between Hyères and Frejus is a gentle, very green range of mountains. On the other side of Frejus begins the Massif de l'Esterel, which is lower and has less vegetation than the Maures Massif and extends all the way to Cannes. This is where the "classical" Côte d'Azur begins, with many beach resorts and a number of small islands off the coast, the Iles de Lerins and, farther west, the Iles d'Hyères, which are highly frequented by tourists. This coastal segment is at the same time a valuable habitat of the animal and plant world.

Figure 18 illustrates, as an example for the coastal region, nature areas between Ollioules and Hyères which are of particular interest from the ecological, faunistic and floristic perspective, which are called *ZNIEFF* areas (the French acronym for: les **Z**ones **N**aturelles d'**I**ntéret **E**cologique, **F**aunistique et **F**loristique). Also, important bird sanctuaries are shown, called *ZICO* areas (**Z**ones **I**mportantes pour la **C**onservation des **O**iseaux). Particularly valuable nature areas (ZNIEFF 1) are on Cap Sicje, a large forest area below La Seyne-sur-Mer; east of Toulon, on Cap Brun; and on the peninsula Presqu'île de Giens, south of Hyères, which is also an important bird habitat; also, the Ile de Porquerolles, east of Presqu'île de Giens, which is classified in its entirety as an important bird habitat. The coastal area La Colle noire south of Le Pradet is ranked among the areas of interest (ZNIEFF 2).

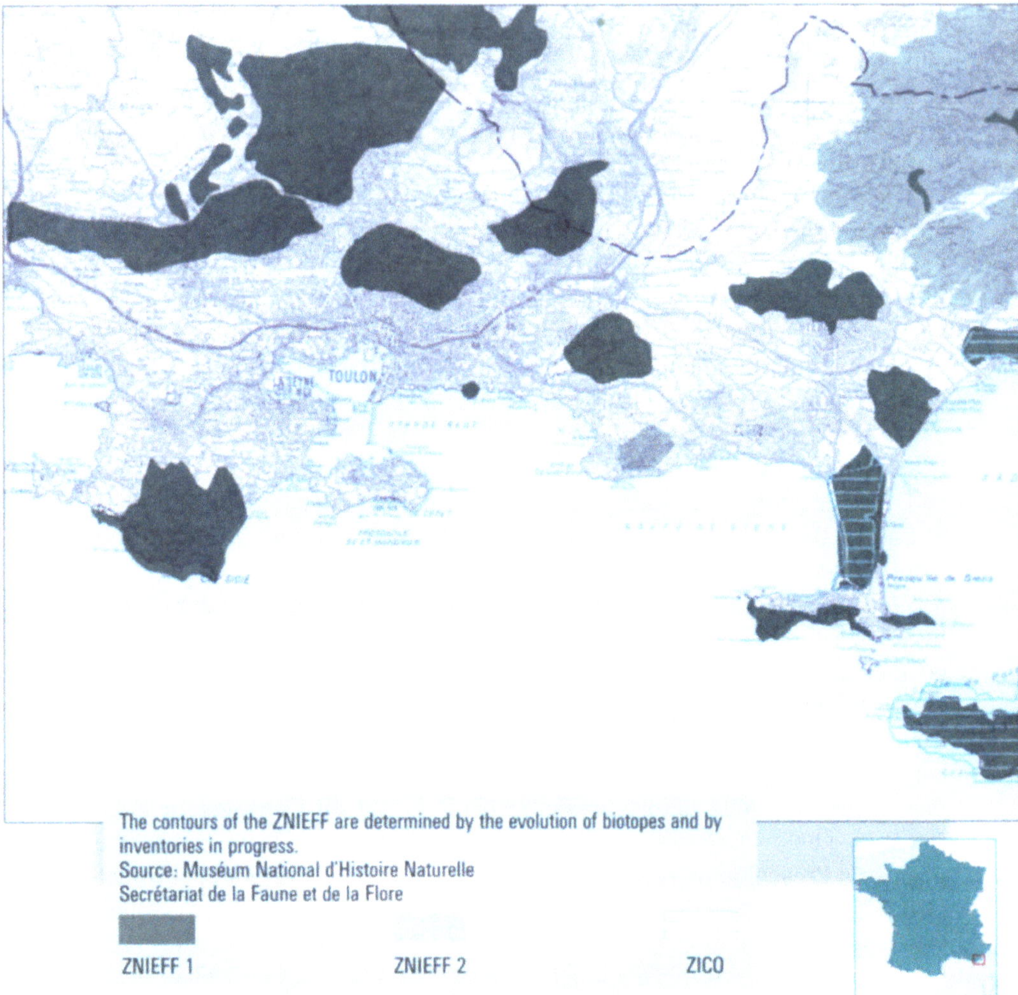

Fig. 18. Nature areas between Ollioules and Hyères particularly worthy of protection. Source: Conservatoire de l`Espace du Littoral 1995, altered.

9.1.3 Conflicting Uses

The share of natural and relatively untouched coastal areas on the Côte d'Azur is steadily declining, because harbours are being expanded and dikes, artificial dunes and beaches are being constructed (cf. DEUTSCHER RAT FÜR LANDESPFLEGE 1984, p. 331–332). *Tourism*, coupled with *massive construction in the dune-rich coastal areas*, has caused a *marked degradation of the coastal dunes* on the Côte d'Azur.

The limits of natural maritime *erosion* were ignored in many places; knowledge of dune dynamics went unheeded. Many coastal towns (La Baule, Arcachon, Saint-Jean-de-Luz) are severely affected by erosion today. This is in part due to *coastal-protection* engineering projects and, where water flowing from the interior maintains the sandbanks, to changed currents or reduced amounts of sediment. In all of these cases the problems stem from a combination of steps geared to protect the cities' strips of coast and the filling in or reclamation of areas to provide additional beach promenades.

Concerning erosion problems, no adequate scientific studies have been conducted in many cases. The responsible authorities often only took action when the erosion process could no longer be ignored.

Today the problem is often combatted simply by filling in the eroded beaches or even completely rebuilding the beach (MIOSSEC 1993, p. 173). This is done by bulldozing the sand elsewhere on the land before the beginning of the season and pumping it onto the beach. After the dirty and finer elements of the material are washed out and carried off by the sea's waves, an attractive white beach remains for the summer tourists.

9.1.4 Solution Strategies

In the last two decades the French government has increasingly attempted to control touristic development and protect the environment. By the late 70's, coastal protection was perceived to be a national objective. The French government has reacted to the negative impact of coastal tourism with a number of laws, e.g., the Law of the Coast (Loi littoral) of Jan. 3, 1986, to regulate the development of tourism. Already in mid-1975, the Conservatoire de l'Espace du Littorale was founded with the aim of protecting sensitive and endangered coastal areas and other inland areas from urban growth and touristic development by buying them up, thus protecting them from further exploitation. The Conservatoire de l'Espace du Littoral is a body of the public administration authorities which is anchored as such in the legal code. With the aid of its right of first refusal and power of eminent domain (in favour of the common weal), areas worth protecting are purchased. The

32 representatives of the Conservatoire de l'Espace du Littoral co-operate closely with the pertinent communities, determining the areas which should be purchased and also developing "rehabilitation plans" (cf. CONSERVATOIRE DU LITTORAL 1995). 20 years ago, on Dec. 23, 1976, a patch of dunes 195 ha in area was acquired in Pas de Calais – the first purchase. By the end of 1990, the Conservatoire de l'Espace already had purchased over 270 properties along the French coasts totalling approx. 36,000 ha. On Jan. 1, 1996, this figure had already risen to 44,334 ha, distributed over 334 areas. Through the purchase of these areas, protection of 622 km of the coastal landscape is secure today. Using the example of the strip of coast between Ollioules and Hyères, *Fig. 19* shows the *already acquired areas (dark blue)* and the *approved areas (light blue)*, i.e., the areas whose sale was still under negotiation as of 1995. The red hatching denotes the areas already protected by the Nature-Conservation Law, such as the Ile de Porquerolles. The broken red line in the Golf de Giens denotes the boundaries of the fishery zone, the RCM (French abbreviation for **R**eserve de **C**hasse **M**aritime).

On the French coast several forms of dune management are being put into effect, all of them geared to keeping the area of the dunes free of all activities, including crossing over the dunes by tourists.

Moreover, attempts are being made to repair damage which has already occurred: after the harbour of Saint-Gilles-Croix-de-Vie was threatened by the degradation and mobility of the dune near La Gerenne, an environmental group had such sensational success when it planted the front side of the dune that its action attracted national attention.

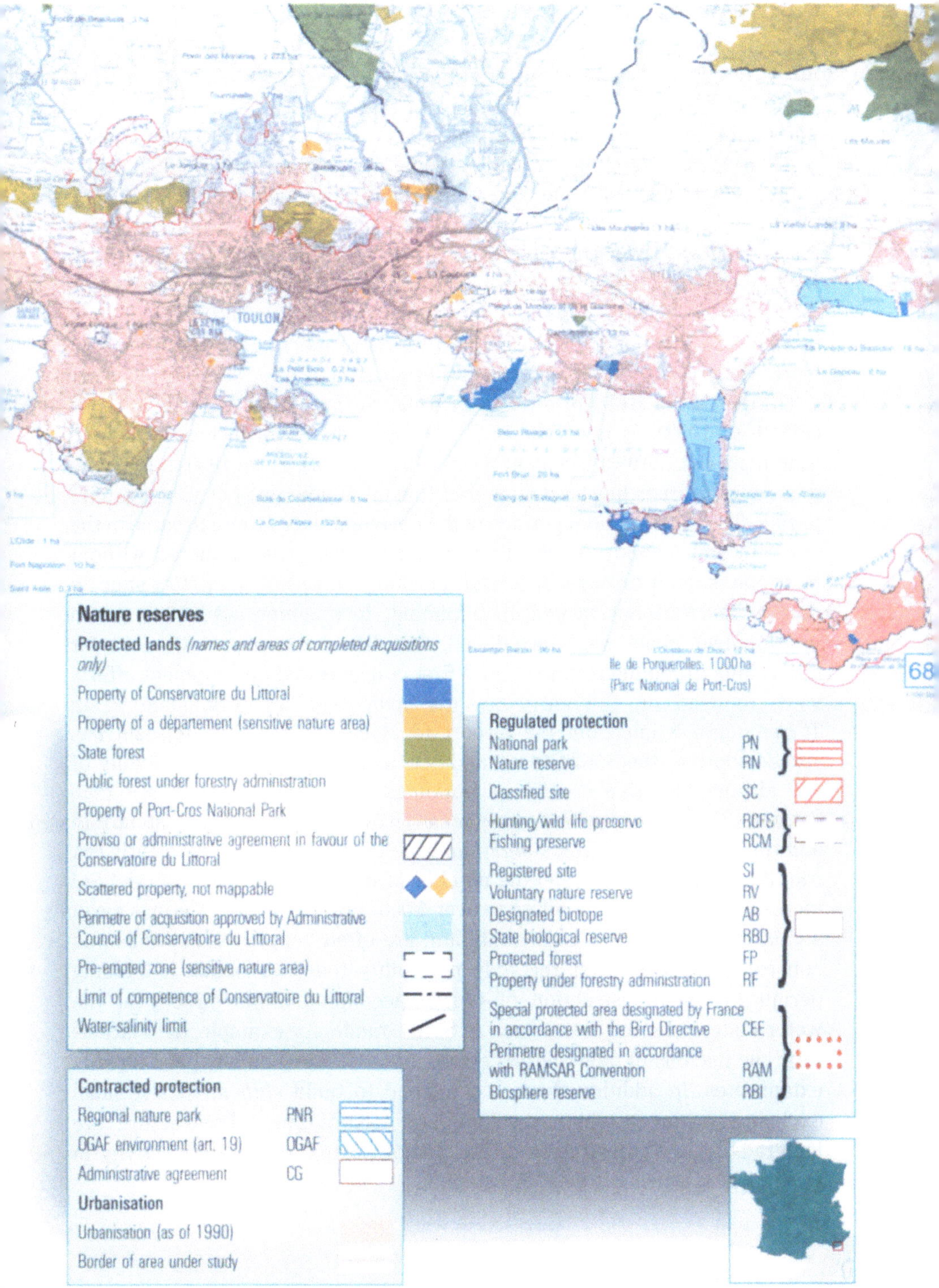

Fig. 19. Nature reserves between Ollioules und Hyères. Source: Conservatoire du Littoral 1995, altered.

9.2 Example: Spanish Coast – Costa del Sol

9.2.1 Touristic Development

The Costa del Sol in southern Spain is one of the most developed coastal regions in the world. Particularly in the last 40 years, the touristic development of Spain has been intensified, and remote fishing and farming villages have been transformed into urban centres concentrated on tourism. After the Spanish Civil War and the Second World War, the new beginning in 1946 was very modest, registering 83,000 visitors/year, but already by 1959, with the opening of the Spanish economy, there were 4 million, in 1975 38 million, in 1985 43 million, in 1992 54 million tourists, and in 1994 the 60-million mark was probably reached for the first time.

On the **Costa del Sol** *alone 6 million visitors were counted in 1993* (FISCHER 1993). By now 50%, i.e., 3,500 km of the Spanish coast is devoted to tourism (McDOWELL et al. in WONG 1993, p. 193). From 1960 to 1990, the number of tourists has grown ten-fold to 6 million people per year (op. cit., p. 190). The biggest boom periods in the development of tourism came in the late 60's and mid-80's. In the Franco Era tourism was promoted without restrictions, as it brought in foreign currency. *Touristic facilities* shot up quickly and without large-scale planning. It was not until 1975, when development plans were introduced, that there was any instrument for controlling urban development. The first concentrated development efforts were focused on the entryways to Spain: the airports, such as in Torremolino, 8 km from the airport in Málaga. Meanwhile tourism has spread south to the Soto Grande and east as far as Narja. In the vicinity of the airport high-rise apartment buildings and condominiums are concentrated, whereas in more remote areas villas are built for the most part. In addition, in the last 10 years *marinas* have been built more prevalently (cf. *Fig. 20)*. From 1980 to 1990 the number of moorings on the Costa del Sol increased from several hundred to over 4,000 (op. cit., p. 204). Marinas mean a gain in prestige for a community and are often coupled with residential construction. They offer the only possibility today to obtain a building permit for the construction of tourist accommodations directly at the water's edge. In some areas, such as Soto Grande, for example, the marinas form the nucleus of new tourist centres; in others they only expand already extant sites. In addition there is the trend to build *club facilities*, small exclusive enclaves where dwelling and leisure-time infrastructure are integrated. They constitute a further major portion of building activity in recent years, especially west of Estepona.

Fig. 20. The Costa del Sol in southern Spain, with location of the largest marinas. Source: McDOWELL in WANG et al. 1993, p. 190, altered.

The economic value of tourism is undisputed: it accounts for 8% of the gross domestic product; of 100 wage earners, 11 are employed in this sector. In 1994 it was estimated that 2.92 billion pesetas would be taken in by the tourism industry. Compared to earlier years, the figure was 2.47 billion pesetas in 1993 and 1.65 billion pesetas in 1988. Spain needs the revenue from tourism to compensate for its negative balance of payments. Tourism brings in nearly double the foreign currency that agriculture does, viz., 31 thousand million DM per year (ABC, Aug. 14, 1994).

In the province of *Málaga (Costa del Sol)* there were 900 overnight stays by foreigners per 100 local inhabitants in the 80's; this statistic was only higher in *Alicante* and *Mallorca.* It is probably higher today. "At that, the popularity of the *Costa del Sol* goes back to the twenties, when well-to-do Englishmen chose the mild winter climate of the southern Andalusian coast for their winter holidays. With the advent of charter air travel, *Torremolinos*, near the airport, became the best known and largest beach resort" (KULINAT, 1986).

After decades of nearly double-digit growth rates, changes in demand in the late 80's due to an expanded supply in other Mediterranean countries and increasing expectations on the part of the tourists led to a drop in turnover for the first time, but also to new touristic developments and other special forms, such as the increasing significance of club tourism mentioned above, the inclusion of *mountain villages and the construction of marinas.*

While all this economic activity was going on, the stress on the environment in the classical environmental media air, water and soil steadily increased. In particular, *contamination of the coastal and inland waters* were caused by the *introduction of untreated household sewage and industrial waste.* Particularly severe was the impact of the natural destruction due to massive tourism in the coastal region. The economic boon that tourism has constituted for Spain as a whole is contrasted by the extreme stress on a small strip of coast.

9.2.2 Particular Natural Features

Spain is a country of relatively high biodiversity, with a high incidence of endemic plants. Apart from the large share of floral elements there is also a wealth of fauna. *The Cordillera bética (and sub-bética) are the areas in Europe and in the western Mediterranean region with the most endemic species. Even if the endemic species are concentrated for the most part in the mountains, they are also found in the deeper areas of Almería and Murcia, as well* (DAVIS et al. 1994, p. 56).

Almería and Murcia are located in a semi-arid to arid climate zone with characteristic, thermo-Mediterranean vegetation. Here shrub communities are found on the steep slopes influenced by the maritime climate, predominantly *Periploca angustifolia or Ziziphus lotus.* The fan palm *(Chamaerops humilis)* is also found in association with *Pistacia lentiscus* or *Rhamnus lycioides. Quercus ilex (subspecies rotundifolia)* formations are the most prevalent climax community. In special locales plants may be found which are not typical elsewhere in the Mediterranean, such as oak forests in the Pyrenees in microclimatic niches with high precipitation and acidic soil. In damper places cork-oak forests are found, and in the southwestern Costa del Sol, near Algeciras, are forests of *Quercus canariensis* (WWF and IUCN 1994).

Special attention should also be paid to the autonomous *Region of Andalusia.* Here there is a plethora of ecosystems, with more than 500 endemic plants. At the same time it is where the last refuges of important but endangered European animal species are located. Andalusia extends over the entire breadth of the south of Spain. The larger portion of its coastline (of 600 km) borders on the Mediterranean; west of Gibraltar it borders on the Atlantic. Central and eastern Andalusia are taken up by the mountain ranges of the Cordillera bética (Upper Andalusia), which has a maximum elevation of 4,478 m in the Sierra Nevada. To the west it is joined by the Guadalquivir Basin (Lower Andalusia). The Atlantic coast *(Costa de la Luz)* is flat and sandy, as opposed to the narrow, steep Mediterranean coast between *Tarifa* and *Cape Gata (Costa del Sol).*

The Guadalquivir Basin and coastal zone have a purely Mediterranean climate, with hot, dry summers. However, the Gulf of *Almería* is highly semi-arid. Whereas in most of the province of Andalusia the Mediterranean vegetation predominates, in the province of *Almería* dry-resistant African floral elements are occasionally found. Of prime importance for fauna is the short bird flight to Africa via the Strait of Gibraltar.

Just west of Gibraltar, at the mouth of the *Guadalquivir,* is the most important wetland and animal reserve in Spain and Europe, the *Coto de Doñana* National Park. Spain has a number of wetlands, 17 of which are designated as protected zones in accordance with the International Ramsar Convention on Wetlands (FISCHER 1993). *Coto de Doñana* is without a doubt the most famous. The national park is basically divided into three habitats with an in part unspoiled natural landscape: wandering dunes near beaches, the highest in Europe; Mediterranean shrub forests and sparse forests with much underbrush and the *Marismas,* broad marshes which are flooded over vast areas every winter (BUFF 1991). "On these loam surfaces, which are covered by a layer of water from October to June, up to a million birds congregate in some years" (RIVERA 1993)

Another wetland area is the *delta of the Ebro,* the largest river in Spain, located in the east. Here the farmers on its banks have been cultivating rice since the middle of the last century. In spite of the use of fertilisers and pesticides, up to 20 million birds live here at times, including ducks, coots, oysterbirds, cormorants and herons.

The Mediterranean sclerophyllous vegetation, because of the ability of its fine branches to dry out and its etheric oils, is extremely vulnerable to fire, but also accordingly adapted. It may be said that a large-scale adaptation of Mediterranean vegetation to the steadily recurrent fire regimes has taken place. But the vegetation makeup has undergone profound changes due to the centuries of human influence, i.e., the introduction of alien plant species. In general the current vegetation, due to the planting of pines and eucalyptus, is much more susceptible to fire, which in 1994 resulted in a loss of 250,000 ha of forests due to fire (EL PAIS, Sept. 22, 1995).

9.2.3 Conflicting Uses

"Just as in the rest of Europe, the destruction of the natural environment is becoming more and more evident on the Iberian Peninsula, too. More and more voices are heard publicly condemning the unscrupulous depletion of the landscape by highway construction and the irresponsible exploitation of natural resources, particularly of water. The often devastating effects of misguided economic, touristic and forestry policies are only now becoming fully evident: toxic industrial waste, soil-depleting eucalyptus plantations." In tourism, in particular, gigantic real-estate and building speculation is accompanied by an aesthetically as well as ecologically questionable *over-development of the coastal landscapes* (FISCHER 1993).

Particularly typical of Spain is the singular phenomenon of *"urbanisation"*. This involves an "urbaniser" (an individual or group) completely planning, developing and constructing a holiday-home and/or apartment-house complex with complete infrastructure, whose individual plots – where almost always construction has already taken place – are then sold. Commercial motivations predominated in many of these projects, resulting in building complexes which were completely alien to the landscape and architecturally inferior. Moreover, the execution of the plan was shoddy. Even with projects which were begun legally and in agreement with the development plan, the park areas set aside by it (a minimum 10% of urbanisation) and the areas reserved for community amenities were also parcelled up and sold. The planning of tourist urbanisations was and is solely influenced by profit motives. There is a wealth of examples of it on the *Costa del Sol*, e.g., *Costa Bella* in *Las Chapas de Marbella*, *Campomijas* in *Mijas*. In many sea resorts and urbanisations there is very dense *high-rise construction*, which is completely untypical of the landscape in design and in such numbers (e.g., *Playamar* near *Torremolinos*). The large-scale urbanisation of *La Manga del Mar Menor* on the *Costa Blanca* on an offshore sandbar, a beach lake (sandbar) 170 sq. km. in area. This urbanisation, designed for 100,000 people, is ecologically as well as aesthetically questionable, as water is difficult to supply. Also, the natural balance of the sandbar is threatened (KULINAT 1986).

Even today *sewage* is discharged untreated into the sea. "On the Spanish coast there are over 200 communities in which sewage treatment is often very inadequate. Thus, according to data of the authorities of Andalusia, over 100,000 people have no access to drainage systems, and 37% of the population must cope with faulty sewers. In addition, for half of all household sewage there is no sewage-treatment facility at all, and for 8% only inadequate ones" (GREENPEACE España 1993).

In the province of *Alicante* one million people live on 160 kilometres of coastline. During the summer months this figure grows to five million. *Benidorm*, 42 kilometres northeast of *Alicante*, has only 34,000 permanent residents, but in the winter 80,000 people live here, and in the summer even 200,000. With the number of residents, the amount of *sewage* also increases, and, due to the overextension of the sewage-treatment plant of *Sierra Helada*, it reaches the sea in an untreated state. This plant "has a capacity of 36,000 m^3/day; but when the number of people there during the summer months increases, the capacity is insufficient for the 48,000 m^3/day then obtaining, and the sewage is discharged untreated directly into the sea. This is done by using a graduated system of pumps pumping it up to 132 m above the sea at Racó de Loix and then dropping it 80 m from the steep coast of the *Sierra Helada* into the sea." The consequences are well known: a rise in nutrient content in the sea causes a disproportionate increase of certain algae and impairment of the natural balance in the sea (GREENPEACE España 1993).

Spain is a country of extreme contrasts: in the northwest there is plenty of water, but towards the southeast it gets scarcer and scarcer. The dryest area is around Almería (Andalusia). For five years now, vast regions of Spain have been suffering from a *drought* the likes of which have only rarely occurred in this century. The "state of the hydrological systems is alarming: (...). Of the available total amount (ground and surface water) all systems but two (which are shared with the provinces of Aragón and Catalonia) are deficient. This situation is entirely typical of the Mediterranean coast" (LA ROCA 1993).

The cause of this is not necessarily the lack of autumnal and hibernian precipitation. It is a result primarily of the water mismanagement in the country: "outdated irrigation methods in agriculture – at 80%, the main consumer – squander half of the amount consumed; in the cities the losses amount to from 25 to 50% due to deficiencies in the water mains. At 1,174 m^3 per capita, Spanish consumption is also far above the European average (726 m^3)" (LA ROCA 1993). The price of water (35 pfennig/m^3) is extremely low (in Germany it is 2.67 DM; ZIMMER 1995).

In some areas, for example Andalusia, water has been rationed for years. Some villages are only supplied by cistern lorries. But the tourists, who bring in the cash, must not notice any of this. "To be sure, there have been water cut-offs in almost all the towns on the *Costa del Sol*, the longest of them in Marbella, where they are only revoked between 6 a.m. and 2 p.m. And the urbanisations at higher elevations, the 'bed castles' consisting of small white houses, for which the water pressure is insufficient, have been supplied for some time now by cistern lorries, as well. But most hotels, bars and restaurants have plentiful reservoirs on the roof or in the cellar, which they fill when the water is turned on. So the only saving this step really achieves is the fact that during the hours the water is turned off nothing can be lost from the leaks in the water mains. In contrast, the more than thirty golf courses on the Costa del Sol are still covered by green grass. So the tourists can still think that the water supply here is inexhaustible" (ZIMMER 1995).

A severe but significant assessment is provided by a column in the news magazine Cambio 16, in which RICO-GODOY writes: "*We have developed a* **cheap-tourism industry** *for whose sake we have ruined the landscape and coasts. The thousands of millions of pesetas which the government must now spend for the farmers' drought compensation must be paid for by the tourism revenues of this season. The day will come when we natives will be swimming in shit while the little water there is goes to the tourist centres. And like a nightmare the day is approaching when we will dress up like tourists so we can take a shower and our children will learn how to play golf so they can put their feet on real grass*" (quoted in ZIMMER 1995).

The government reacted to the emergency with immediate aid for the farmers. Pessimistic predictions claimed that in 1995, 50,000 small farmers would give up in southern Spain. In a new large-scale project called the "National Hydrological Plan", which has been debated in Parliament since 1993, the entire water reserves of Spain are to be surveyed so that they can be fairly distributed over the entire country. At that, it is economic interests which are behind the calculations of future water demand: the expansion of agriculture as an export sector, the utility companies and tourism as a source of foreign currency. Increased "surplus" water resources are to be conducted to economically strong but ecologically destroyed regions, against which there is much opposition. Critics of the plan reason that it would be better to try all possibilities for making water consumption more efficient and adapt the predictable requirements of development plans to the amount of water available, e.g., in large-scale tourism projects (LA ROCA 1993; ZIMMER 1995).

9.2.4 Solution Strategies

That something must be done to combat the one-sided concentration of Spanish tourism on the coasts has been undisputed among Spanish tourism experts for a long time. The catchword is *"diversification"* of the masses of tourists.

"There are enough resources and possibilities for this in the interior, but too little money was set aside for it in the 80's. An attempt was made to arouse interest in winter sports in the Pyrenees, the *Sierra de Guadarrama* and in the *Sierra Nevada* (...). Other ideas were to expand the thermal spas and religious, sports and cultural tourism. First attempts at offering holidays at farms were made in Catalonia. All in all, however, the diversification of Spanish tourism is only beginning. Moreover, it will not be successful until this objective has been fully recognised and accepted by politicians, the sciences and the business community" (KULINAT 1986).

After becoming conscious of its wealth of flora and fauna, increased awareness of the special national characteristics of the historically evolved, natural and cultural landscape structures and nature areas and encouraged by becoming a member of the European Union in 1986, Spain has undergone a tremendous development in nature conservation. If Spanish nature-conservation policy, as in most European countries, was traditionally limited to setting aside certain picturesque landscapes, it has managed, thanks to the effects of the European nature-conservation and regional policies and to the transferral of authority to the regions *(Comunidades Autónomas)*, to expand its protection policy considerably. In particular, the adoption in April, 1989 of a law providing the framework for nature conservation based on the European drafts for the FFH Habitat Directive opened the field for nature protection at the national level (SCHMIDT 1995).

With the *nature-protection framework law* Spain is aiming at (SCHMIDT 1995):

- a complete revamping of the classification of protected-area types,
- support for the European nature-conservation strategy with the nature-conservation programme Natura 2000 (the programme provides for protection of about one fourth of the area of Spain) and
- comprehensive nature-protection planning along regulatory as well as development lines, with emphasis on protected-area planning.

The various planning schemes provide for a combination of nature protection and environmentally compatible development of the *eco-desarollo* "based on the socio-economic analysis of the plan area and its environs". According to SCHMIDT (1995), "it must be seen as a model for the implementability of European nature-protection goals in economically depressed areas while adhering to the principle of equal value of living conditions". The premise of Spanish nature-conservation policy is "the preservation of mixed models of use, similar to the example of the French 'soft' nature parks" (SCHMIDT 1995).

"This nature-conservation planning comprises national and regional regulations for protecting natural resources in the form of guidelines: the natural-resource plans preceding the certification of protected areas (*Plan de Ordenación de Recursos Naturales* – PORN) and the plans guiding the use of the protected areas (*Plan Rector de Uso y Gestión* – PRUG). The planning of the protected areas is put into action by annual action programmes. The natural-resource plans worked out by the nature-conservation authorities provide a framework for the intended certification of the protected areas within which the subsequent nature-conservation plans must be confined." The nature-conservation planning of the law of 1989 thus provides, at least for the rural areas ($^1/_4$ of Spain's total area), an alternative plan which looks to the future (SCHMIDT 1995).

The nature-conservation law has thus far only been complemented by laws of implementation in four regions: *Asturia, Castilla y León, Murcía and Andalucía.* But only the last-named region has adopted significant expansion of the national guidelines, as set forth in the nature-conservation law of Andalusia of February, 1989. With this regional law Andalusia was able to expand its network of protected areas considerably, from 5.1 to 15.7%. This ensures that the nature-conservation authorities have planning authority for these areas. However, SCHMIDT (1995) criticises that this strategy will lead to a lack of direction in protected-area policy, as the participation and co-operation of the directly affected parties is virtually excluded.

In Andalusia today there are over 22 nature parks (parques naturales) with a total area of 1,412,830 ha, which equals 16.2% of the region's total area *(cf. Fig. 21:* nature reserves in Andalusia). Some 1.5 million people live in the NPs. The Andalusian environmental authority responsible for the NPs *(Agencia de Medio Ambiente)*, founded in 1984, is striving to develop a touristic programme which is compatible with nature conservation. This form of nature tourism is geared to open up new sources of income for the populace. Apart from that there is one national park *(parque nacional)* and various nature reserves *(reserva natural)* similar to the German *naturschutzgebiete* (protected areas) and landscape-protection areas *(paraje natural).*

Fig. 21. Nature reserves in Andalusia. Source: Dirección General de Turismo, Junta de Andalucia 1995, altered.

Of the 22 nature parks (NPs), four are on the sea. These are the NP *Entorno de Donaña*, the NP *Bahía de Cádiz*, the NP *Acantilado y Pinar de Barbete and the NP Cabo de Gata-Nijar.* On the Mediterranean as well as on the Atlantic coasts various *reservas y parajes naturales* have also been established.

Andalusia is here following an ambitious course. With the establishment and erection of facilities in 22 nature parks it is realising a diversification of its touristic programme by developing tourism in nature and rural areas. Rural and nature tourism are being expressly promoted by the government, as stated in a number of publications. The interior, which is characterised by high unemployment (35%), a weak economy and general lack of perspective, needs to be developed. At the same time, pressure needs to be taken off the coast, which by now is saturated by development. *The mistakes made in the development of coastal tourism should not be repeated. Rural, agricultural, nature and ecotourism are the catchwords* (GOMEZ et al. 1992).

With the only recently inaugurated project GESNATUR, the Andalusian environmental authority is promoting local private initiatives in the nature parks to work out unified quality criteria for visitor centres, information offices, camping grounds, rural accommodations, forest huts, recreational activities and other facilities for public use. A communication plan has been begun to increase the awareness of the nature parks' existence and inform about their wealth of natural and cultural assets as well as possibilities for investing in the nature parks.

Along with the *Coto de Doñana* National Park under the administration of the state Institute for the Preservation of Nature (ICONA), an additional nature park was established a few years ago (which is within the authority of the Andalusian regional government) whose 50,000 ha serve as a buffer zone for the National Park against anthropogenic influences. "Protests against a planned hotel to be built there moved the Andalusian regional government and the enterprises involved with the hotel project to sign a common declaration of intent in August, 1993 to look for another site – albeit just as big a one, which is still to be determined – which would not be quite as near to the National Park" (FISCHER 1993).

9.3 Example: Turkey – Southern Coast

9.3.1 Touristic Development

As early as the 60's, Turkey began to expand its tourism sector: the establishment of the Ministry for Tourism and Information in 1963 marked the beginning of the promotion of foreign tourism as a national policy. In the same year the first five-year plan was inaugurated, which provided for a planned development of the tourism sector. During the first plan period (1963–1967) the number of tourists from abroad increased to 23.5% compared to earlier years and along with it a 9% increase in revenues from tourism. During this period the state made the greatest portion of tourism investments for the construction of infrastructure, hotels and holiday villages, some of which are run by the Ministry itself (cf. ALTAN in: BECHMANN/FAHRENHORST 1993).

In the second period (1968–1973) the planning for expansion was continued, but now the emphasis was placed on the promotion of private investment. To prevent bad investments and to increase the efficiency of investments, priority areas for touristic development were stipulated: the Mediterranean coast, the southern Aegean, Göreme *(cf. Fig. 22)*.

Fig. 22. Priority area for tourism in Turkey. Source: ALTAN 1991 in: BECHMANN/FAHRENHORST 1993, p. 139, altered.

For the planning and promotion of these priority areas a planning directorate was established, subordinate to the State Bureau of Planning (DPT) whose task it was to draw up touristic-development plans. In the third plan period (1974–1978), development in the tourism sector was principally concentrated on the priority areas. This was based on the assumption that the number of foreign tourists would steadily increase. Whereas in 1971 only 926,000 foreign tourists came to Turkey, by 1982 as many as 3.3 million and by 1985 5.3 million were expected.

For the further expansion of the tourism sector the third and fourth five-year plans formulated the following principles and tasks:

- steering investment and subsidies in the direction of mass tourism,
- promotion of private investment and inclusion of foreign capital,
- easing of accessibility to loans, combining of domestic and foreign tourism in the same facilities,
- expansion and improvement of special training for tourism personnel.

Despite all these steps, however, foreign tourism in Turkey had by the end of the 70's no significant growth rate to show for itself. The main reasons for this were the weakness of the domestic capital market, lack of interest on the part of domestic investors, domestic unrest and the economic policies at the time, which disproportionately promoted the manufacturing and investment-goods sectors, but not the tourism sector, which is a source of foreign currency. Also, the quality of the extant accommodations was below the standard expected by foreign tourists. The few hotels in Antalya or on the Aegean were unable to satisfy the expectations of the tourists. Airports, flight connections and the rest of the transportation network were not geared to tourism on a large scale, and other touristic infrastructure was totally lacking.

With the aid of the World Bank, three detailed plans for touristic development were drawn up by the state bureaus of planning. The choice of the three development areas of Köycegiz, Antalya and Side, in which tourism was granted absolute priority, was made according to the following criteria:

- sparsely populated coastal area with a correspondingly unspoiled landscape and clean environment,
- climatic conditions conducive to a long season,
- attractive topograpy and vegetation (forests),
- regional and/or locally typical, historically significant sites,
- interesting architecture and urbanistic structures of the surrounding cities and villages,
- extensive infrastructure (airports, yacht marinas, roads; utilities/waste-disposal facilities; electricity, water, sewage etc.)

(BECHMANN/FAHRENHORST et al. 1993b, p. 8).

The re-development package of the International Monetary Fund constituted a crucial impetus in promoting tourism in Turkey. From 1980 to 1983 the tourism sector gained ground in Turkey. This much time was needed to gain international trust and to make tourism attractive for investors and tour operators. In 1983 the number of tourists rose compared to the previous year by 16%; by 1987 the number of foreign visitors rose by 47% to 2.4 million. To accomodate such hordes of tourists, the capacity of hotel beds was increased between 1983 and 1985 to 107,000. Apart from the hotel beds licensed by the Ministry, 150,000 more beds are available in pensions approved by local authorities and in other accommodations (cf. BECHMANN et al. 1993b). Below, two of the areas of touristic development will be discussed in detail: Köycegiz and South Antalya.

A. The Köycegiz Development Area. The Köycegiz development area comprises the area around the city of the same name on the southern Turkish coast between Marmaris and Fethiye. The six-kilometres-long strip of coast is just as much a part of the development plan as the ancient city of Kaunos, with its hot springs, as well as four inland lakes, one of which, Lake Köycegiz, is the largest freshwater lake in the entire Mediterranean region *(cf. Fig. 23)*.

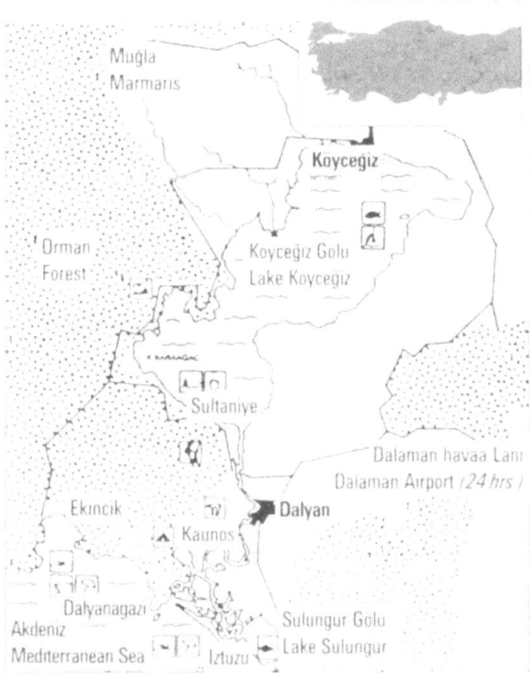

Fig. 23. The Köycegiz/Dalyan area. Source: SCHERFOSE/YÜCEL 1988, p. 33, altered.

In 1978 the construction of *infrastructure* was begun; one year later, the first foundations of tourist facilities were laid. By the planned end of the development steps in 1990, a total capacity of 6,000 beds was to be achieved in *several hotels and holiday villages.* The Kaunos Beach Hotel in the area of the 4.5-km-long mouth of the Dalyan Delta, whose cornerstone was laid in 1987 on the beach of Iztuzu, was to account for 1,200 of these beds alone *(cf. Fig. 24).* 51% of the project was in the hands of the Turkish business group KAVARA. The IFA Hotel und Touristik AG of Duisburg, Germany and the German Financing Association for Investment in Developing Countries jointly held the remaining shares.

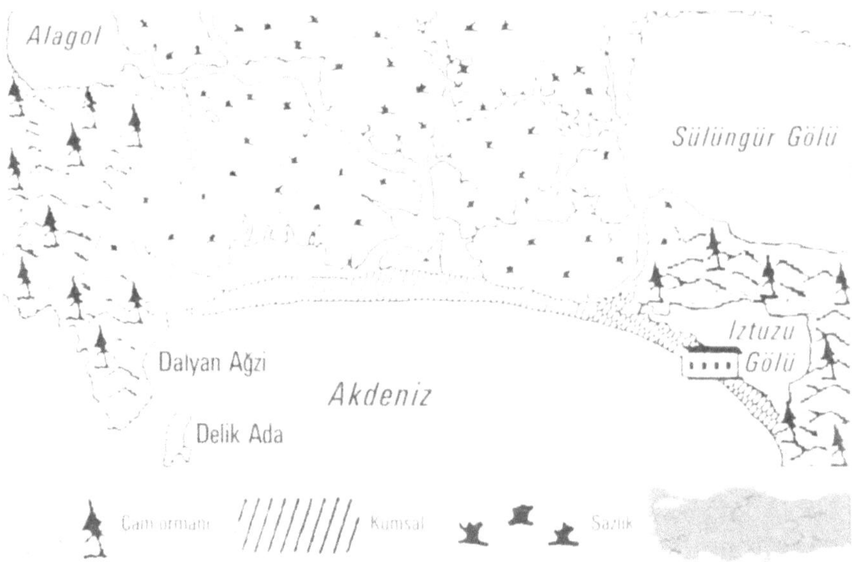

Fig. 24. Planned site of the Kaunos Beach Hotel. Source: SCHERFOSE/YÜCEL 1988, p. 33, altered.

9.3.2 Particular Natural Features (A)

The nearly untouched ecosystems of the sandy beach, the brackish delta, the lagoons and Lake Köycegiz have a number of floral and faunal particularities. The swampy, reedy area is surrounded by steep mountain slopes covered with pine. Towards the sea it is bordered by a sandbank about 4 km long. The entire area surrounds a body of water and reed belt approx. 95 sq. km. in area.

On the shores of Lake Köycegiz is the greatest concentration of the liquidambar species endemic in Turkey and on Rhodes *(Liquidamber orientalis)*. Originally from Asia Minor, the liquidambar was cultivated in Southern Europe because of the liquid resin of its inner bark (cf. BECHMANN/FAHRENHORST et al. 1993a, pp. 15ff.).

Apart from its being home to over 150 different bird species, the area is a resting place for thousands of migratory birds and the habitat of the otter and a number of freshwater turtles. A study conducted by the WWF indicates a high species diversity, with 7 fish, 4 amphibian, 9 reptile and 155 bird species, as well as countless plant species which have been not yet precisely charted.

The Dalyan Delta is considered to be one of the four important breeding grounds of the sea turtle *Caretta caretta*. The sea turtle, which is included in Appendix I of the Washington Convention on Endangered Species among the highest-priority species to be protected, uses the sandbank near the Bay of Iztuzu for laying its eggs. The "great or loggerhead turtle", which can be up to 1,2 m long and weigh 100 kg, is phylogenetically one of the three oldest surviving animal species.

Because of the uniqueness and pecularities of its landscape and from the standpoint of landscape protection as well as for its function as a habitat for a number of different species, some of which are threatened and endangered, the area around Dalyan should be classified as being particularly worthy of protection.

9.3.3 Conflicting Uses (A)

Every year some 17,000 tourists are brought through the reeds to the ancient city of Kaunos in flat-bottomed boats. In addition, during the main tourist season, about 20,000 tourists come from Marmaris to Dalyan, Kaunos and the beach at Iztuzu (daily in the main season, cf. BECHMANN/FAHRENHORST et al. 1993a).

The conflict resulting from the strong *tourist frequency in the nature area* is illustrated very clearly using the example of the sea turtle *Caretta*

caretta. Whereas just a few decades ago 30,000 specimens of *Caretta caretta* were alive, a count conducted in 1986 revealed a decline in stock to approx. 3,000 animals.

In 1979, 300 *sea turtles* still made their way to the sandbank near Iztuzu Bay to lay their eggs; by 1986 only 58 sea turtles were counted.

The mature female returns to her own birthplace every three years to lay her eggs in the sandy beach there. Damage from tourism stems from the possible destruction of the incubation cavity by sunshades stuck into the sand, the plundering of nests by dogs or by people taking the eggs home as souvenirs. Furthermore, the activities on the beach can result in older animals not digging any incubation cavities or that the young cannot dig their way out after hatching. The young hatch after 55 days and must reach the sea immediately after they have dug their way out: various bird species and crabs reduce the survival rate of the turtles on their nocturnal way to the sea to approx. 10% of the eggs laid. As the young turtles in their instinctive march to the sea presumably use the light reflecting on the seawaves to guide them, they are distracted by artificial light on the beach. Moreover, deep footprints or tyre tracks are an insuperable obstacle for the turtles, which are only five centimetres long. Once they are in the sea, the young are subject to another danger: their soft shells are vulnerable to the propellors of *tourist boats*.

On the beach at Dalyan conservationists (among them the WWF) prevented the construction of a hotel in 1987, to ensure that the sea turtle *Caretta caretta* could lay its eggs without being disturbed. The news of this success was reported by all the media for weeks. At the time Dalyan was an unspectacular fishing village. "Since then, attracted by the many press reports, more tourists have been making their pilgrimage to this paradise regained every year, where they reverently view the foundation of the halted hotel construction and hope to see one of the newly hatched baby turtles (...)" (cf. TROTTNO 1992). Over 300,000 tourists came from the nearby hotels in 1991 and took their tour operators up on the offer to enjoy a day in "unspoiled nature". The loggerhead turtle has become a trademark – on the neon signs of the restaurants, the T-shirts of the boat captains bringing the tourists to the beach and in the souvenir shops. In the mid-80's there were only simple pensions in Dalyan; today there are already 3,000 beds, and they also meet high quality standards (op. cit.).

B. The South Antalya Development Area. The touristic-development area is located on the border of the province of Antalya, to the west of the provincial capital Antalya. It comprises a strip of coast from 3 to 10 km wide and 80 km long. As early as 1965, the area west of Antalya was stipulated by the Turkish cabinet as a touristic-development area; by 1974 the first legally binding development plan (DP) on a scale of 1:25,000 had been submitted, which was approved in 1976 and even today (with several changes of development stipulations, *cf. Fig. 25)* constitutes the basis for touristic development in the area.

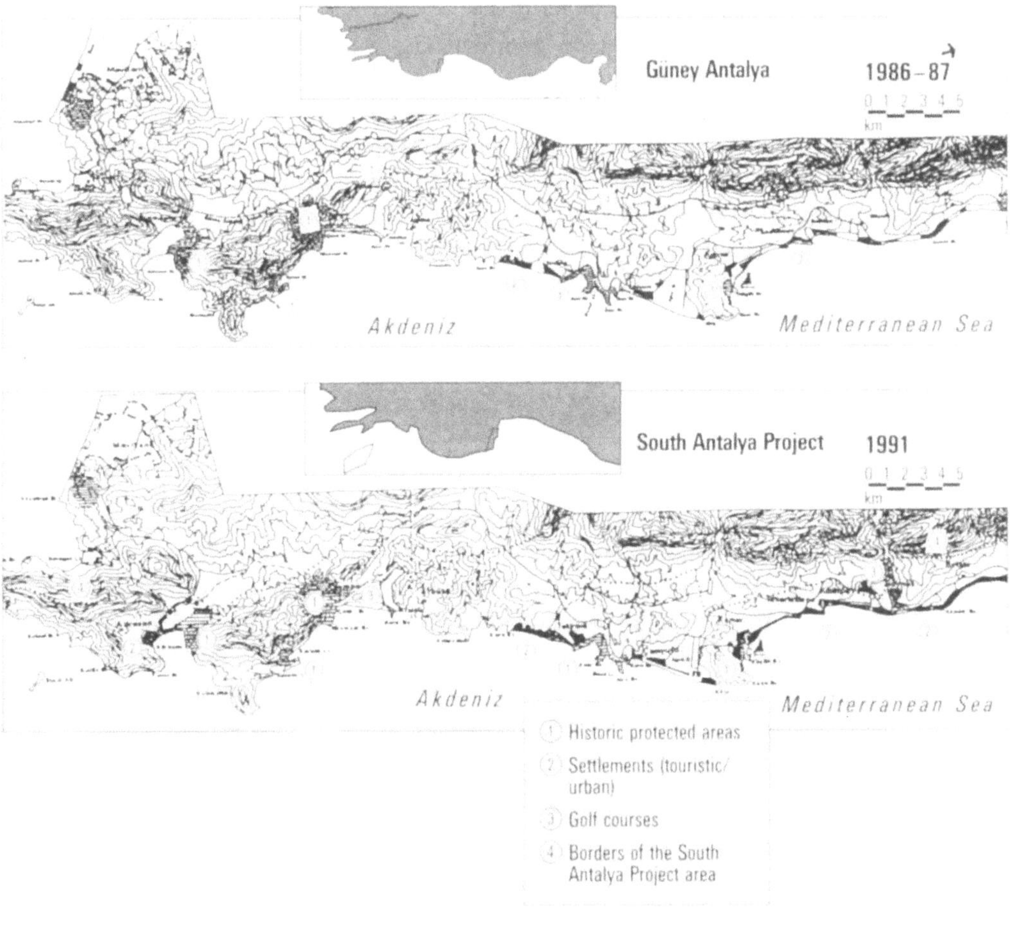

Fig. 25. Zoning changes in the South Antalya Project area; plans of 1986/87 and 1991. Source: BECHMANN/FAHRENHORST et al. 1993b, altered.

The DP provided for a capacity of 25,000 beds and had the goal of preparing the area for international mass tourism. Due to its geographical location, the village of Kemer (population 600 at the time) was to be expanded to form a centre which would assume the service functions for the urban and touristic development of the area. According to the development plan of 1975, Kemer was to have 10,000 inhabitants by 1996.

The original calculation of bed capacity was based on the "beach capacity". The average beach width was assumed to be 40 m. The beach required per person or bed was estimated to be 16 sq. m. This meant that 0.4 m of beach on the coast must be available per bed. With this calculation the bed capacity was computed to be 21,070 for the length of the beach of the entire area, to which a bed capacity of 2,313 campers and daytime tourists was added (BECHMANN/FAHRENHORST et al. 1993b, p. 38). After numerous changes of the DP, the bed capacity, so carefully calculated at the time, has by today been increased to 69,000 beds (op. cit., p. 25).

9.3.2 Particular Natural Features (B)

The entire area is located within the Olimpos-Beydaglari National Park, the planning of which was begun in 1970 because of its potential for tourism, historical value, unique scenic beauty and interesting nature. The touristic-development areas on the coast and the productive mountain forests were taken out of the National Park, however – an area amounting to 50% of the original National Park.

The touristic-development area is characterised by a narrow strip of coast behind which the steeply climbing Taurus Mountains rise to an average height of 3,000 m. The sandy beaches in this area average from 100 to 250 m in breadth and are bounded every 10 to 20 km by foothills of the mountains jutting out into the sea.

The most important factor for touristic development in this area is the climate, which is very favourable due to the shield towards the north posed by the Taurus Mountains; it belongs to the thermo-Mediterranean climate zone.

About 81% of the area is taken up by forests; 19% of the project area is under cultivation (mainly citrus fruits, maize; sesame on the slopes). The natural forest type of the area is conifer forest, the main species being *Pinus brutia*, which in the Mediterranean climate grows from the seacoast up to 1,100 to 1,200 metres above sea level. As the forests have been well protected up to now, they are relatively dense. They generally grow right down to the coast.

9.3.3 Conflicting Uses (B)

The increase in bed capacity has primarily been responsible for the drastic impacts on the touristic-development area, which are:

1. problems in providing infrastructure
2. problems in providing water
3. problems in removing garbage
4. problems with the urban development of the town of Kemer.

All the utility facilities were designed in accordance with the original plan of 1975. As the actual number of beds is much higher, as described above, a great many bottlenecks result which are additionally exacerbated by shifts in the quality class of touristic investments. In the higher quality classes the consumption of water and energy automatically goes up, which in turn leads to increases in the amount of sewage and refuse (cf. op. cit., p. 49).

In order to compensate for the water deficit, the tourist facilities have already resorted today to finding their own water supply, i.e., with the permission of the State Water Authority (DSI) deep wells are being dug. But this causes the layer containing salt water to rise *(cf. Fig. 26)*, so that it reaches the roots of the trees growing in the coastal area and damages them. In most of the areas devoted to tourism these problems are already occurring.

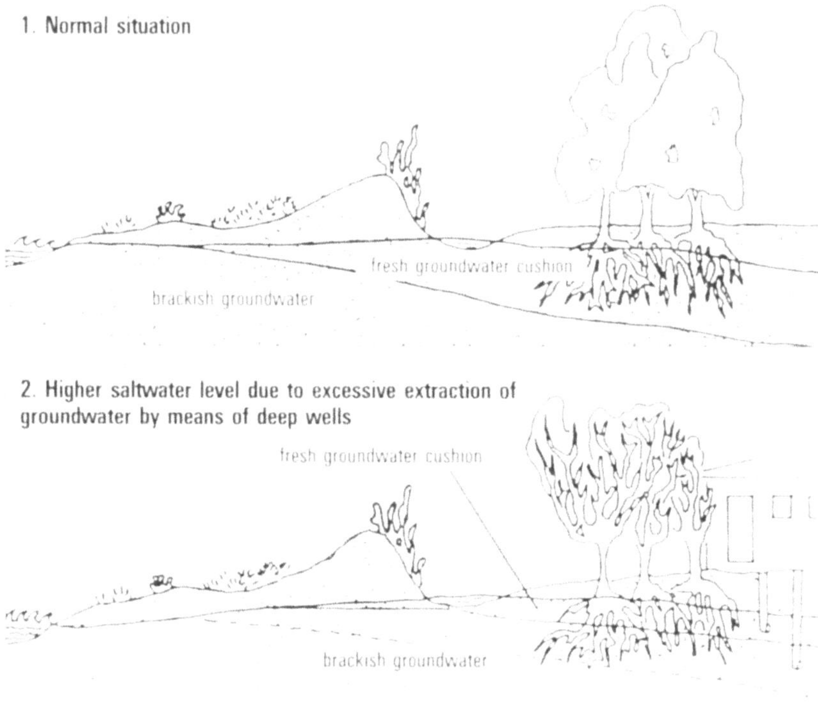

Fig. 26. Rising saltwater level on the coast. Source: ALTAN 1992 in: BECHMANN/FAHRENHORST 1993, p. 51, altered.

According to the 1975 planning, the population of Kemer was to be about 10,000 by 1996. But Kemer already had more inhabitants by 1990, and the municipal authorities estimate that by 1996, 20,000 people will be living in Kemer. Originally only certain areas were stipulated as residential or such areas where new houses could be built, as the character of the town was to be preserved. Large portions of the southern district of town were placed under protection for historic reasons. In the meantime, however, further areas have been opened to construction, some of them in the protected area, not least due to pressure from investors.

All of the services for tourism are run by companies from outside the area, so that the native populace has little share in them. The formerly idyllic village of Kemer has by today developed into an "artificial" city devoted exclusively to tourism (cf. BECHMANN/FAHRENHORST et al. 1993b).

9.3.4 Solution Strategies

The Turkish government had set itself the goal of avoiding the mistakes other Mediterranean countries had made with their tourism – Spain, for example – and taking comprehensive protective steps for the environment. The fact is, however, that Turkey has destroyed its coastal ecosystems far more extensively in the past 15 years than Spain did in the last 40 (SALMAN 1994).

In 1982 environmental protection was added to the Turkish constitution as a human right and state goal. But environmental protection and coastal experts have scarcely been included in coastal planning; recommendations by national and international experts for the sustainable development of coastal areas have largely been ignored (cf. op. cit., p. 3).

At the legislative level, several individual provisions have been made regarding tourism, such as the environmental law (no. 2872, 1982), the national-park law (no. 2873, 1983) and that governing coasts (no. 3086, 1984). In addition, there is a law for the protection of natural and cultural objects and the law governing construction. However, these legislative steps must not conceal the fact that at the level of political decision-making in Turkey those development plans are being given preference which render comprehensive environmental and nature protection impossible. Decrees for their realisation are absent in most cases (BECHMANN/FAHRENHORST et al. 1988). There is no unified control at the central level, no environmental planning on a departmental plane nor an integrated, regionally relevant environmental planning. Protective legislation such as the environment or national-park laws are faulty; laws regulating access and use (e.g., that on the coasts) contain few protective measures.

Since the reform of public administration of 1984, which transferred much responsibility to local authorities, environmental standpoints have had to yield to local interests. At the national level, too, the General Authority for Environmental Questions appears to be relatively insignificant. In addition, there is a widespread lack of personnel in the institutions qualified to perform control functions. Whereas instruments to generate incentive, e.g., financial support for environmentally friendly technologies, only exist to a limited degree, the penalties for infractions of the law in this field are very slight and ineffective (cf. BECHMANN/FAHRENHORST et al. 1993a, p. 45).

9.4 Example: Germany – Baltic Coast

9.4.1 Touristic Development

For many decades tourism has proven to be a major source of income on the North-German coast. Important tourist areas on the southern Baltic in Germany are the resorts in Schleswig-Holstein, Fehmarn Island, the coast of Mecklenburg-Vorpommern and the island world off the shallow coast, especially Rügen (cf. LIEDL et al. 1992).

The Baltic coast of Germany is in particular a popular destination for the Germans themselves and thus constitutes a typical example for *domestic-tourism development* and its impact on the coast.

As to where the tourists come from, surveys for Schleswig-Holstein have shown that they come mainly from the northern regions of Germany. The vacationers in Mecklenburg-Vorpommern, in contrast, come from all over Germany (cf. op. cit.).

In the Bay of Lübeck, for example, the share of foreign tourists (up to the political turnabout) was only 3% in 1989; on Rügen the figure was 6%, at any rate, which is explained by exchanges with other communist countries in the former GDR (cf. HELFER 1993). On the West German coast of the Baltic it is the Scandinavians who, often only in transit, make up the small percentage of foreign visitors.

Whereas the Baltic coast was only one of many destinations for West German tourists – in 1989 only 2% took their holiday on the West German Baltic coast – the East German coast was for many citizens of the GDR far and away the main holiday destination (over 33% in 1989). One third of this number was concentrated on Rügen.

Figure 27 illustrates the development of overnight-stay statistics from 1950 to 1989 in the Rostock region and the County (Landkreis) of Rügen; *Fig. 28* provides the development of overnight-stay statistics on Rügen from 1957–1994. Due to different survey methods, the statistics on overnight stays in the Baltic resorts of Schleswig-Holstein before 1960 and after 1980 are not comparable to the East German figures: whereas the East German statistics include all communities, those on the Baltic resorts of Schleswig-Holstein do not inlcude communities with fewer than 5,000 overnight stays per year.

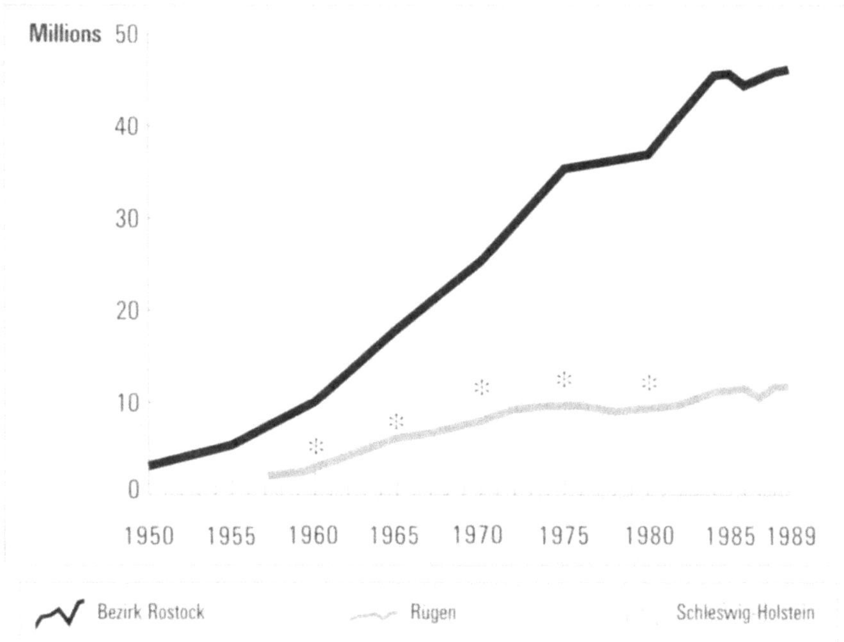

Fig. 27. Comparison of overnight-stay statistics for the East and West German Baltic coasts. Source: Statistisches Landesamt Schleswig-Holstein, Urlaubsstatistik des Bezirks Rostock, in: HELFER 1993, p. 35, altered.

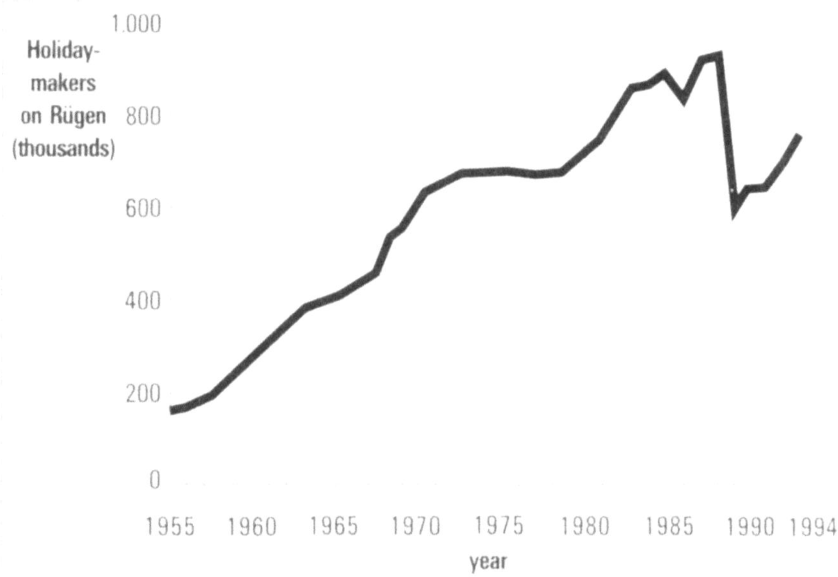

Fig. 28. Development of tourist statistics on Rügen from 1957 to 1994. Source: HELFER 1993, LANDKREIS RÜGEN 1993 and 1995, in: SPITTLER 1996, p. 30, altered.

This illustration clearly shows that touristic development on both coasts, after a steady increase up to 1975, stagnated in the late 70's. In the Rostock region and on Rügen there is a renewed sharp growth of overnight-stay figures from the outset of the 80's.

In 1989 the number of tourists on Rügen reached a high mark of 915,290 people. In the first season after the monetary union of 1990, the number of holidaymakers dropped to 586,000.

"Including the daytime and weekend holidaymakers not counted in the statistics, the total number of people visiting Rügen in 1988 is estimated at 2 to 2.5 million (Landkreis Rügen 1993). *Far more severe was the decline in overnight-stays, which dropped from 26 million in 1989 to a mere 3 million in 1990. As visitors spent far less time on Rügen, the average stay declined from 14 to 2–3 days. By 1994, the number of holidaymakers reached 750,000, and overnight stays increased to 3.5 million, and the average stay rose to 5–6 days* (Landkreis Rügen 1995). *An estimated 750,000 more people visited the island during the day in 1994, bringing the total to 1.5 million. (...) After the sharp drop in 1989, it appears that the number of visitors to Rügen will again attain a high level"* (SPITTLER 1996, p. 29).

9.4.2 Particular Natural Features: the Example Rügen

Rügen, with an area of 962 km^2, is the largest island of the southern Baltic coastal region (cf. KNAPP et al. 1988). The islands of Ummanz, Liebes, Wührens, Mährens, Heuwiese and Lieschow Peninsula belong to it. Only the island of Ummanz, which is connected to the Isle of Rügen by a bridge, is populated, as is the Lieschow Peninsula; also, in the western portion of Rügen itself, there are but few settlements. The Strela Sound, a channel formed by melting ice in the Pleistocene, and the Greifswald Shallows separate Rügen from the continent.

"The landscape structure of today reflects many thousand years of history: it is characterised by broad ground moraines on southern and western Rügen, clefted terminal moraines in the eastern portion of the island and the chalk landscape of the Jasmund Peninsula, which juts out like a block. Basins formed by glacier tongues and former dead-ice fields are taken up by extensive marshy lowlands; young sandbars connect the island's Pleistocene nuclei with one another" (KNAPP et al. 1992, p. 62). A total of seven landscape types can be distinguished on Rügen, as shown in the following *Fig. 29*.

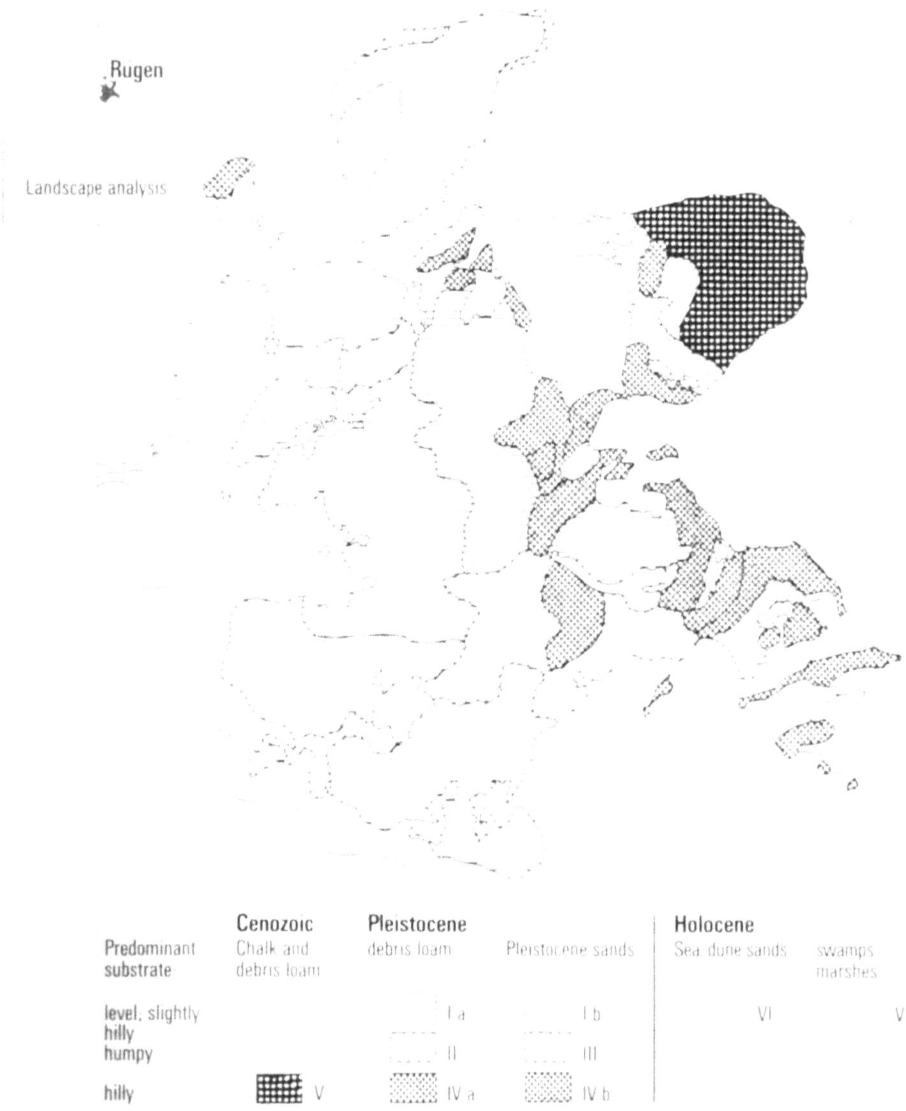

Fig. 29. Landscape analysis of the Isle of Rügen. Source: KNAPP et al. 1992, p. 60, altered.

Due to its craggy profile and the very fertile soil in places (marl from glacial drift), a very diverse vegetation formed there. In many places the soil is under cultivation. For this reason, the forest, which once covered large areas on the island, has been reduced to numerous scattered stands of trees. Among the particularities are the forests of chalk-beech trees on the Jasmund Peninsula (SCHNEIDER 1994).

In the moraines and glacier-tongue basins are small marshy hollows and extensive marshy lowlands. On the end-moraine sandy slopes, eskers, dunes on cliff edges and the sandbanks, lean sandgrasses predominate. On the marl slopes of the terminal moraines semi-arid grasses and dry bushes took root (KLEINKE 1992, p. 56).

Typical of Rügen are also the salty waters of the shallows which have only one or several narrow connections to the open sea and thus limited water exchange with the Baltic. The shallow-water areas are the most biologically active and ecologically crucial zones of the near-coastal waters. They are refuge areas for many rare species of the animal and plant worlds, such as the sea eagle, osprey and cormorant. For resting migratory birds such as cranes they are important overnight stops.

Several gull and duck species, terns and waders find nesting places on Rügen. The avifauna is markedly diverse.

"The diversity of the landscape, the interesting flora and fauna and particularly the intensive interpenetration of land and sea constitute the particular attractiveness of Rügen. None of the 319 settlements is more than 8 km away from the coast. Steep coasts rising up 100 m high with gleaming white chalk rock and broad sandy beaches stretching for kilometres, dark beech and light pine forests, green meadows, fields bordered by poppies and cornflowers, peaceful bays and charming beach resorts prevent any boredom from setting in. The richly varied craggy appearance, with numerous lookout points, is a further enhancement to getting to know the island's landscape. Has Rügen lost any of its attractiveness in the last 200 years? The number of its tourists has at any rate been steadily increasing (...)" (HELFER 1993, p. 39).

The elementary precondition for an attractive island tourism is usually adequate beaches. Experience shows that the greatest attraction is constituted by broad sandy beaches on the open sea. In this regard nature has endowed Rügen richly. Rügen has a total coastline of 575 km. On Rügen's outer coast there are 39 designated beaches and bathing spots with a total length of 47,950 m. On the shallows coast there are 41 designated beaches and bathing spots with an available length of 13,160 m. A further 5,200 m of beaches on the outer coast and 1,350 m on the shallows were prohibited to bathing in 1993 for reasons of hygiene and a further 3,040 m on the outer coast and 250 m on the shallows for reasons of nature conservation. So Rügen has a total of over 61.11 km of beaches of unlimited availability (cf. op. cit.).

The impressive steep chalk coast of Jasmund was the main attraction on Rügen in the early days of tourism, which go back to Romanticism. Today the white chalk cliffs of the steep coast continue to attract the eye, even though vacationers are more strongly concentrated on the beaches. Precisely in the North German landscape, where it is otherwise so flat, the towering walls of rock do not fail to make an effect. But above all, the attractiveness of the widely varying island landscape for tourism should not be underestimated.

9.4.3 Conflicting Uses

The growing number of visitors is leading to increasing environmental stress. According to a study by HELFER (1993), tourism and leisure-time infrastructure on the Isle of Rügen have had a damaging impact particularly on the *forests protecting the coast.*

"The growing stream of holidaymakers and motor vehicles led on the one hand to expansion of the resorts and the construction of large building complexes, e.g., in Binz, Sellin, Baabe, to the expansion of the network of roads (...), to a change in function of small settlements and farms near the coast and the laying out of large camping grounds with extensive permanent installations (bungalows, trailers, utilities, sanitary installations); on the other hand this extremely strong concentration of holidaymakers exerts considerable pressure on the coastal landscape: destruction of the vegetation cover and erosion caused by the effects of automotive traffic, trampling and reclining on the steep coasts, on beach walls and dunes, eutrophication and littering of the beaches and excursion sites, stress from garbage and sewage (...)" (SCHMIDT et al. 1975 in: KNAPP et al. 1986).

Pine forests on sandy soil are particularly sensitive to damage from trampling and driving, erosion and ground solidification. Because of the close proximity to the sea, numerous camping grounds have been established in them, a significant part of which are permanent camping facilities. *Figure 30* shows the damaged segments of the coastal-protective forest, totalling about 6 km. Even simple camping grounds, but particularly the many car parks on the beaches or in the free landscape, constitute massive disturbances of the ecological balance. The damage consists primarily of:

- exposing, damaging and destroying tree roots,
- damage to the trunk and branches,
- destruction of ground vegetation from trampling and driving on it,
- erosion as a result of destruction of the soil cover by digging drainage ditches etc.,
- contamination of the soil by dumping of wash water etc.,
- moderate to strong thinning out of the coastal-protective forest for car parks, paths and clearings for activities as well as the severe limitation and in part complete stillstand of forest replanting, leading to a disproportionate aging of forest stock etc.

In the environs of near-coastal tourist facilities, an anthropogenically exacerbated coastal erosion is often found. In some places strongly eroding unauthorised paths from cliff to beach which have not been closed off (or not effectively) have been found to exist. On some coastal segments there is already clear *evidence of overdevelopment by tourist and leisure-time infrastructure.* In addition, *sports and leisure-time activities* are very destructive to the environment. *"This is especially true of boaters and surfers cavorting near or in the protected areas and of hikers and horse-back riders who leave the trails"* (LIEDL et al. 1992). Holiday-bungalow areas or permanent camping grounds and holiday camps take up entire coastal segments in some places and are not accessible to the public. Add to this a *visual and aesthetic impairment of the landscape,* which can stem from tourist facilities when they are in exposed places and are not adapted to the structures which have evolved there. Modern functional buildings are a particular eyesore, but also camping grounds or holiday-bungalow plots with insufficient screening. *Figure 31* shows the visual/aesthetic impairments of Rügen's island landscape, particularly in the especially sensitive coastal area.

Fig. 30. Damage to landscape caused by tourism. Source: HELFER 1993, p. 109, altered.

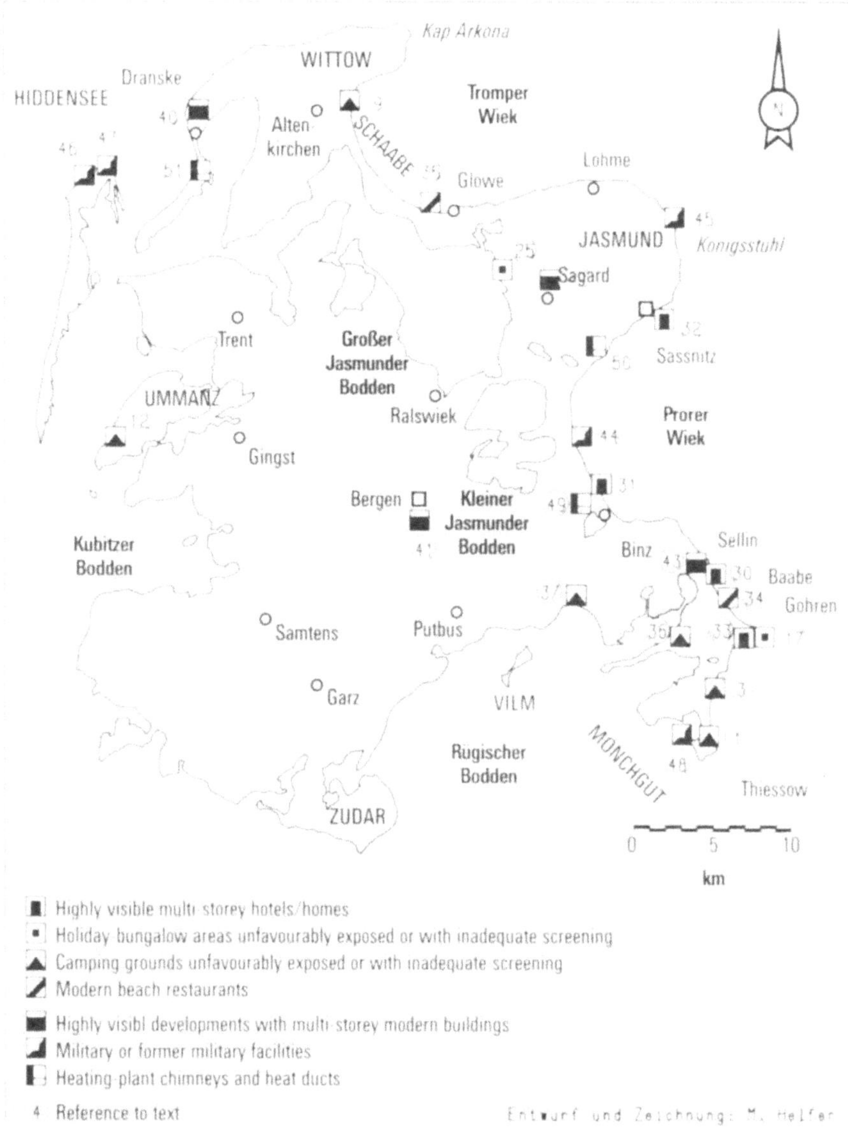

Fig. 31. Visual/aesthetic impairment of landscape. Source: HELFER 1993, p. 111, altered.

9.4.4 Solution Strategies

Already in 1990 a first draft for a structural plan for Rügen and environs had been drawn up. In the revised version of May, 1991 it forms the basis for working out a development plan at the county and regional level. Among these are development guidelines for tourism and recreation, but also those regarding nature conservation and landscape management.

In the area of tourism and recreation the general goal of the fundamental restructuring is "to develop the area for the improvement of social, economic and environmental living conditions for the populace of and visitors to the county *(of Rügen)*. Of particular importance for this goal are:

- the securing of jobs,
- the preservation and development of the characteristic structure of settlements and the typical landscape,
- the preservation and development of natural values and functions."

In particular, a considerable increase in the quality of the tourism sector is planned, in which the scenic and cultural particularities of Rügen and its environs will serve to develop a characteristic range of touristic and recreational attractions which is of particular interest because of its very uniqueness.

For the coastal area the "zone ordinance" of the structural plan makes the following stipulations:

- In the entire shore zone, which is at least 100 m wide, tourist use must take place only on the basis of the building plans;
- a strip of shoreline 100 m wide is to be kept free of construction;
- on the long, drawn-out beaches of the Schaabe and between Göhren and Thiessow, the hinterland is to be kept free of construction and car parks as far as possible, to protect the landscape; access is to be secured by means of public transportation;
- the character of the resorts as spas and medicinal baths should be built up again and developed in connection with improving the environmental situation;
- camping grounds are to be removed from the forests protecting the coast;
- solid buildings which have been built without pertinent regulations should be examined as to their compatibility with the structure of settlements and landscape and removed, if need be;
- aquatic sports, particularly pleasure boats, must be offered suitable facilities, without constructing unprofitable overcapacities, e.g., at moorings; in protected areas and shore zones aquatic sports are to be restricted;
- sports which are detrimental to nature-compatible recreation in the landscape are to be prohibited.

For the area of nature conservation and landscape management the "Plan on Large-scale Protected Areas" is significant.

As early as December, 1989, a plan on large-scale protected areas for the GDR was discussed for the first time, at the urging of several citizen-action groups. The transitional government of the GDR placed 14 large landscapes under legal protection on Sept. 12, 1990, shortly before unification with the Federal Republic; 23 other areas were temporarily secured. In the Unification Treaty between the two Germanys, the protected status of these areas was confirmed. The protected areas stipulated for Rügen are shown in *Fig. 32*. Newly established were:

- the rather small (30 sq. km.) Jasmund National Park with the chalk coast and Stubnitz Forest;
- the Vorpommersche Boddenlandschaft National Park, 805 sq. km. in area, comprising the island of Hiddensee and also Bug Peninsula and the largest part of the west coast of Rügen as well as the nearby peninsulas of Fischland, Darss and Zingst and the shallows between them, whose 768 sq. km. of area make up the largest part;
- and the Southeast Rügen biosphere reserve, 230 sq. km. in area (111 sq. km. of which are water);
- an area for an East Rügen Nature Park was temporarily secured; its management will be established in the coming years. With that, half of the County of Rügen is subject to the regulations of nature conservation (cf. HELFER 1993).

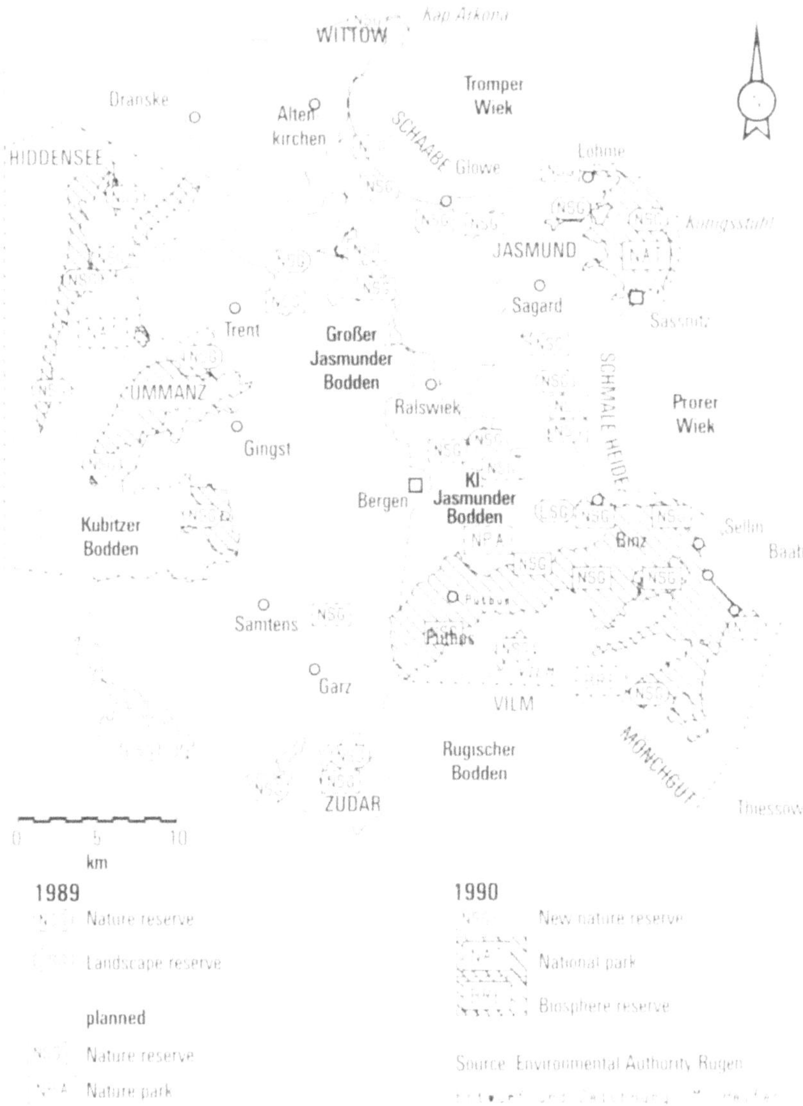

Fig. 32. Comparison of nature reserves, 1989 and 1990. Source: HELFER 1993, p. 179, altered.

On the basis of the development guidelines a landscape framework plan is being drawn up. For areas worth protecting which are not included in the protected areas a protection, development and management plan is being worked out. Currently, on the initiative of a leading Rügen County official, and with the support of many communities and environmentalists, the idea is being discussed to declare 96% of Rügen a nature park. Thus, the development of a "soft tourism, coupled with cautious development of other economic sectors" is being planned (TKALEC 1995).

9.5 Example: Germany – North Sea: Schleswig-Holstein Wadden Sea

9.5.1 Touristic Development

Tourism is one of the most important activities in the Schleswig-Holstein Wadden Sea National Park and thus one of the main sources of income for the people in this region, as well. Tourism in the Wadden Sea has increased appreciably in recent decades. More leisure time, coupled with an improved standard of living and greater mobility of many people, have resulted in the Wadden Sea being frequented for recreational purposes throughout the year. With about 16 million overnight stays and 11 million excursion tourists, who account for revenues of some 1.4 thousand million marks combined and thus for nearly 20% of the entire real net output of the region, tourism is of crucial importance to the population's livelihood. Moreover, it is the only growth sector in the region (cf. WWF-Deutschland 1995). A strong concentration on the islands (over half of the houses on the islands are used for tourist accommodations) and the holiday centres on shore (particularly St. Peter-Ording and Büsum), with sometimes extreme economic dependence on tourism and a marked seasonality, are the characteristics of tourism in the National Park area. Tourism entails an infrastructural development of the coastal area and a seasonal as well as spatial expansion of various leisure-time activities such as aquatic sports, sport aviation, hiking and horseback riding in the mudflats and a more intensive touristic freqency of the sandy beaches, dunes and salt meadows.

9.5.2 Particular Natural Features

The German North Sea coast, including its many islands, is over 1,000 km long. Its area was reformed by glaciers several times. During the last Ice Age great amounts of sand and pebbles were transported to the coast from the northeast, which explains the regional petrography and the predominant sediments (loose sediments, beach and dune sand, marshland and clay). The greatest part of the flat North Sea coast consists of sand and marshland extending in places more than 20 km into the interior. The marshy coast of the shore was formed during the course of the last millenia by the tidal cycle and periodic flooding.

The offshore islands protect the area of the marshland. The water level changes with the tides by several metres, particularly during the winter storms, with west winds reaching force 12 (KELLETAT 1993).

"The Wadden Sea is an ecosystem found nowhere else in the world, with widely varying habitat types and of great significance for the nearby ecosystems. It is a habitat for some 100 species of nesting and resting birds, a rearing area for many fish species of the North Sea, habitat for about 4,000 smaller animal species in the shallow near-coastal waters and the salt meadows and for about 100 species of flowering plants which are dependent on the habitats of salt meadow, dune and sandy beach" (Ökosystemforschung Wattenmeer 1/92, p. 7).

9.5.3 Conflicting Uses

On the North Sea coast there is a multitude of competing forms of use. According to a report on ecosystem research on the Wadden Sea ("Ökosystemforschung Wattenmeer" 1/92), tourism is one of the dominant ones in the East Frisian region. During the tourist season, the number of holidaymakers arriving there is from ten to a hundred times the native populace. The *leisure-time activities* and *additional emissions (sewage)* of these masses of tourists are a seasonal cause of disruption and stress and occur at the very time of greatest biological activity. In the shore area in front of the dikes, the establishment of *camping grounds* and *bathing beaches* has resulted in considerable *areas being taken up by tourism.* Whereas for example the populace of the island of Sylt have always built their houses a minimum of 250 m away from the sea, tourist buildings were positioned as close to the beach or cliff as possible. The buildings were thus directly exposed to the winter floodings and erosion processes and were destroyed, sometimes after only a few years.

"The disturbances and stress on the coast stemming directly from recreational activities can be described qualitatively, but thus far have not been sufficiently identified and quantitatively measured, so that – with regard to the significance of tourism in this region – much research remains to be done" (op. cit., p. 31). The following direct impacts have been observed thus far:

- *damage to dunes and vegetation* from trampling, with subsequent aeolic removal and increased erosion sensitivity of the dune front to sea surf,
- other use and sealing off of natural areas by *hotels, roads, apartment houses, harbours/marinas etc.*,
- *generation of garbage* and *pollution from sewage*,
- transformation of the natural coastal system into an artificially structured strip of coast due to engineering projects required for protection from erosion,
- *disturbance of the habitats* of birds, seals and fish.

After the first beach walls were erected with private funds, the first publicly financed steps to protect dunes and beaches were initiated in 1865. By now the marshland is protected by dikes totalling 1000 km in length. Without the enormous investments in connection with the development of tourism there would be no necessity to protect the dunes and coast to this extent. All man-made elements of the coastal system have ecological consequences. The dikes separate different areas of the tidal coast from each other and thus prevent the exchange of water, nutrients and sediments. After the conventional technical coastal-protection solutions proved to be very costly, in part ineffective and often unattractive, as well, with regard to tourism on the coast, a relatively natural method has been put into use since the early 50's. Since 1951 beaches have been artificially restored. As this has little impact on the continuing erosion processes, it is cheap and effective. To preserve the coastline, about 1 million cubic metres of sand must be piled on the beaches every 5 to 7 years.

9.5.4 Solution Strategies

As to the ecological significance and desirability of protecting the wadden coast, there is international agreement today: the National Park is a recognised protected area within the framework of the Ramsar Convention. The UNESCO recognised the Schleswig-Holstein Wadden Sea as a biosphere reserve in 1990 within the framework of the MAB (Man and Biosphere) Programme.

For the protection and preservation of the Wadden Sea, the German Federal Government created the Schleswig-Holstein, the Lower Saxon and the Hamburg National Parks of the Wadden Sea. In addition, the Wadden Sea was chosen as a second research area, after Berchtesgaden, for the ecosystem research programme of the Federal Ministry of the Environment, Nature Conservation and Reactor Safety (BMU). Within the scope of trilateral government negotiations between the three bordering states,

Germany, Denmark and the Netherlands, a common protection plan is being developed based on ecological monitoring.

The task of protecting the Wadden Sea ecosystem presupposes knowledge of the complex structures and functions of the system. As in many areas there is still no thorough knowledge of the natural relationships and the consequences of human intervention, research on the ecosystem Wadden Sea is being supported by Schleswig-Holstein and Lower Saxony, the BMU and the BMFT together. The results of this interdisciplinary research project will provide the basis for the management and protection plans in the national parks. With the simultaneous establishment of lasting environmental observation, statements on environmental quality will be gained which will provide the framework for environmental-policy decisions.

The state government of Schleswig-Holstein, with its tourism plan of 1990, has opted for the "soft tourism" development strategy. Soft tourism is to become the "trademark" of the holiday region. The state government is planning to compile an environmental stock-taking for resorts (ecological success assessment) (cf. KERN 1995).

In the tourism policy of the state of Lower Saxony and of the communities of the coastal region there is agreement today that nature is a tourist attraction which is gaining more and more importance and that for economic reasons alone the strategy of tourism must be geared to preserving nature (cf. NIEDERSÄCHSISCHES FREMDENVERKEHRSPROGRAMM in: SCHARPF 1989, p. 263). The Ministry of Economics is supporting projects with a specific view to promoting environmental compatibility and social acceptance. Within the framework of this economic support the Economic Support Fund – Ecology – "Tourism" Directive (Ecofund) is an exception, as it serves tourism exclusively.

As a consequence of the ecosystem research of the Wadden Sea, the Bureau of National Parks, the bordering communities, the county with its institutions and the nature-conservation association in care of the Hamburg Hallig have formed a working group to discuss steps for overseeing tourism on the Hamburg Hallig. The first step would be a toll barrier which would make every visitor think about whether driving to the Hallig in his car is worth the money. As an attractive alternative, going there by bicycle or on foot continue to be free of charge. Besides, the limitation of car-park capacity on the Hallig would restrict the growing automotive traffic (cf. ÖKOSYSTEMFORSCHUNG SCHLESWIG-HOLSTEINISCHES WATTENMEER, p. 108).

On the Isle of Föhr the town of Wyk is taking particular steps to find solutions to the severe volume of traffic on the island stemming from tourists. Already in 1982 a system of one-way streets was devised to make access to certain areas more difficult for cars. In a central area in the interior a ban on driving at night was introduced; on downtown streets no-stopping zones were set up and 30-km/h speed limits imposed (cf. TU BERLIN 1989).

9.6 Example: Ireland – Northwest Coast

9.6.1 Touristic Development

In recent decades a great number of *tourist activities* have been concentrated on the coast of Ireland. Many small holiday towns, particularly in the vicinity of large urban centres, have been expanded by construction, particularly of *holiday homes*. Portballintrae in County Antrim, for example, consists to about 25% of holiday homes (CARTER et al. 1993, p. 221), the majority of which are used by residents of Belfast. This is an adverse trend for the local residents, as real-estate prices have sharply risen as a result. In Northern Ireland, in particular, *house trailers* are left there as *holiday homes* and only used on weekends and during holidays.

Another important and also touristic use is golf. Although *golf* has been played on the coastal dunes for a century, the building of many new courses in recent years (e.g., in Portmarnock in Co. Dublin, Portstewart in Co. Londonderry and Clifden in Co. Galway) have markedly changed the coast. Also, bays and estuaries are increasingly being used as sites for *marinas* (e.g., near Arklow and Greystones in Co. Wicklow, Bray in Co. Dublin, Ardglass in Co. Down or Castletown in Co. Kerry), even if the development in Ireland cannot yet be compared with that in Wales and the south of England.

The Irish coast is today almost completely accessible; even cliff sites are furnished with marked paths. The number of foreign tourists on the coasts is difficult to assess, however, as there is a lack of meaningful, differentiated statistics (domestic/foreign tourism; city/coastal tourism etc.). According to CARTER et al. (1993) it should be assumed that a large portion of tourism on the coasts stems from the local populace and their guests (friends and relatives).

The most frequently visited tourist attraction in Ireland is the Giants Causeway in County Antrim, with over 250,000 people per year (CARTER et al. 1993, p. 221). The entire coastal segment in County Antrim attracts over one million visitors every year (op. cit.). Most visitors sleep in house trailers, however, which are often placed in sensitive coastal-dune segments.

How many people visit which beach can likewise not be said unequivocally, due to the lack of data. At best, it may be said that Brittas Bay in County Wicklow probably attracts about 200,000 people per year, and Portstewart in County Derry at least 100,000, as here the number of cars on the *beach car park* were counted only during 10 weeks of the summer (CARTER et al. 1993, p. 221).

Many sites only attract great numbers of visitors for a short time, on hot summer weekends, and are deserted the rest of the year. In Tyrell in County Down, for example, over 2,000 cars were counted on a single summer (weekend) day (op. cit.). The area accessing such beaches is very small: most visitors come from less than 50 km away (op. cit.).

9.6.2 Particular Natural Features

The Irish coast is approx. 3,200 km long and is composed of a series of bays, estuaries, beaches, dunes and cliffs. But there are significant differences between the east and west coasts. In the west the combination of heavy ocean surf and strong west winds has led to a spread of dunes and broad sandy surfaces often stretching inland for kilometres. On the more placid east coast it is primarily the erosion products of the cliffs from which the sediments on the Irish Sea originate. The flat sandy beaches east of the Wicklow Mountains between Bray and Arklow are some of the most beautiful on the island.

Fig. 33. *(left)* The coasts of Ireland. Source: CARTER et al. 1993, p. 213, altered.

The island is situated in an area of mild southwesterly winds and the Gulf Stream. This results in a relatively temperate climate with no appreciable temperature differences within the island. The average air temperature in January/February is in the range 4–7°C and in July/August, 14–16°C (ELVERT 1994, p. 19). Precipitation is distributed over the island by the predominant westerly winds. The west has higher average precipitation, however, up to 3,000 mm per year, whereas the east of Ireland seldom has more than 750 mm per year. In Ireland there are relatively few days without precipitation; the sunniest months are July and August, with an average period of sunshine of 6 hours (ELVERT 1994, p. 19).

On the island of Ireland there is a relatively low species diversity compared to other European countries. As a consequence of the separation from the European continental shelf after the last Ice Age, a large portion of the plant and animal species died out.

The island was originally covered largely by oaks, which by now have been cleared except for a few pristine forests. In regions with calcareous soil there was also a rich stock of ash, hazel and yews.

The reforestation programmes have hardly paid the original vegetation any heed, and have above all been concentrated on rapidly growing conifers (Sitka spruce and Douglas fir, Scotch pine, stone and other pines, larch, Norway spruce etc.).

In regions with poor run-off in the central lowland plain, mountain bogs have formed due to the plentiful precipitation. There numerous peat-moss, heather and reed species occur. Killkarny, Co. Cork, and Glenngarif, Co. Kerry, are two particularly interesting areas. Here there are extremely oceanic and even tropical bryophytes and lichen species. In the "Burren", in County Clare, there are also Mediterranean plants alongside the arctic-alpine species which have survived the last Ice Age.

Of the 380 species of the birds living in freedom observed in Ireland, over 135 nest on the island. In addition, birds establish their winter quarters here, most of them coming from Greenland and Iceland. For example, 75% of all the Greenland white-fronted geese spend the winter in Ireland. The great attraction Ireland has for birds is partly explained by the plethora of inland bodies of water which provide some species with particularly favourable conditions for forming colonies

Whereas geese, swans, oystercatchers and many duck species make their home primarily at river mouths, auks, gulls and kestrels settle on the Atlantic coast, where there is a wealth of marine animals.

A great portion of the worldwide storm-petrel and gannet populations nests on the south and west coasts of Ireland. The species diversity of mammals is basically similar to that in the temperate climate zones of Europe.

9.6.3 Conflicting Uses

The human influence on the Irish coastal system has thus far been relatively limited compared to Central European coastal areas, say. There is no great sediment transferral nor are there strong tides, so negative impacts are limited to small areas. These areas include dunes, in particular. Here there are a multitude of human impacts which have resulted in the *destabilisation of dunes, the loss of marshes and lagoons and the removal of coastal sand.* A large share of the natural diversity has already been lost, and this trend has been accelerated in recent years by the *intensive growth of construction* on the coast. Moreover, the large numbers of tourists, the easy accessibility to the coast and the great variety of *leisure-time activities* impair the ecological state of the coast. Most evident is the ecological and geomorphological *damage to the coastal dunes.* In Portstewart, for example, the dune ecosystem completely broke down between 1950 and 1980 due to high pressure from human sources (CARTER et al. 1993, p. 222). In the meantime, protective and management steps have been taken to alleviate this pressure and restore habitats. However, they have not had the desired effect, as the goal of stabilisation can solely be achieved by stabilising the vegetation.

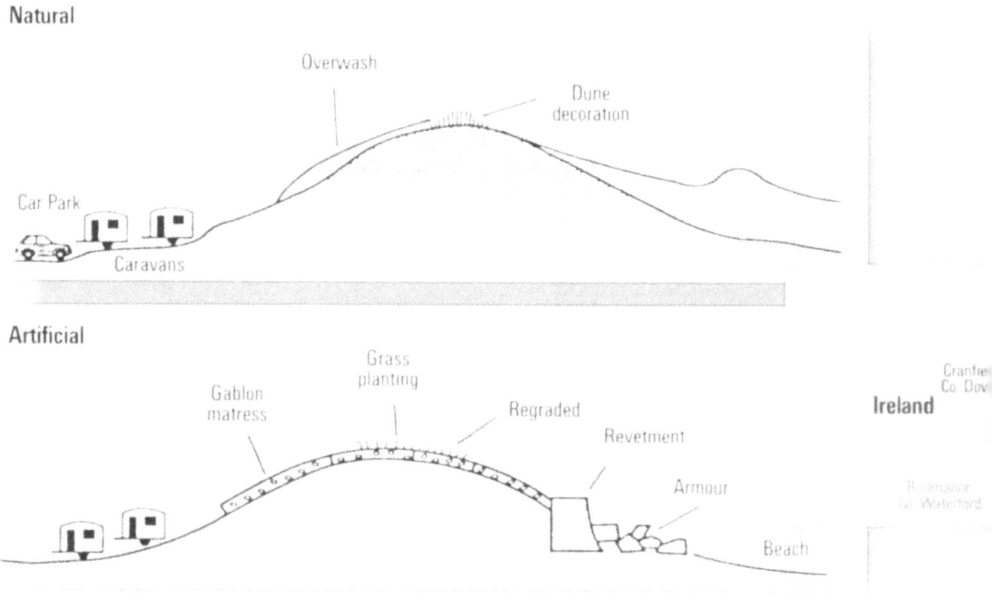

Fig. 34. Use of the former Cranfield Dune (Co. Down) as a site for camping grounds. Source: CARTER et al. 1993, p. 217, altered.

9.6.4 Solution Strategies

Perhaps the best example for dune protection in Ireland is the Murlogh Dunes, a National Nature Reserve in County Down. It was acquired by the National Trust in 1960 and thus saved from reforestation. The dunes are divided into zones of varying use intensity; some areas are inaccessible, and elsewhere the number of visitors is limited. In this way ecosystem protection and recreational use can be reconciled with one another.

The future of the Irish coast depends on future planning and policy decisions. In particular, the rise in sea level due to global warming must be addressed and protective measures be taken. Thus far, neither the Republic of Ireland nor Northern Ireland have assumed a clear and coherent posture on coastal protection. EU directives, such as the bathing-water directive of 1975, have thus far contributed more to the confusion than to clearing up the situation. Planning is often limited to individual sites and "case-by-case" decisions. Instead, supraordinated framework plans such as a strategic coast plan must be drawn up. For damaging impacts, such as house trailers, holiday homes, marinas and leisure-time sport centres co-ordinated planning should be worked out. Uncontrolled growth has caused great environmental damage up to now.

10 Strategies for Achieving Sustainable Tourism in Coastal Regions

"Sustainable" Tourism. The notion of an environmentally and socially compatible tourism, which is today widely referred to as "sustainable" tourism, implies also the improvement of the economic livelihood of people and at the same time the long-term securing of the natural foundations of life. The concept of sustainability means that a "development geared to long-term compatibility" should be strived for, a "lasting development without overexploiting natural resources and without destroying the basis for existence" (quoted from TÖPFER 1993 in: KERN 1995). According to SCHARPF (1994, unpublished) "sustainable tourism" aims at achieving a development which enables a constant or growing tourist demand to be met at a site with a lower environmental burden or constantly high environmental quality.

Procedure. The instruments mentioned herein offer potential approaches for implementing sustainable touristic-development strategies. The structural presentation of the descriptions is adapted from SUMMERER (1995). The instruments described will be of regulatory, planning and monitoring and economic nature, including informal ones, and also integrating instruments and strategies of the European Union for coastal protection. A consideration of the existing solution strategies which analyses their effects (in the form of efficacy studies) does not yet exist, to our knowledge. The following presentation of the instruments for eliminating or reducing touristic stress must therefore be done in largely descriptive rather than critically analytical fashion. Inasmuch as comments and critical remarks as to the efficacy of the instruments were available, they have been included in the description.

Instruments. At the European level there is at present no formal international agreement directly concerning the relationship of nature conservation and tourism. However, solution strategies relevant to sustainable tourism are touched on in international programmes for the management and protection of coastal ecosystems, international and national planning strategies for the protection of biodiversity, directives of the Council of the European Union and the states, legally binding regulations (as part of the development and nature-conservation plans of the states) and recommendations of various European organisations and initiatives (such as those of the European Union for Coastal Conservation [EUCC], or the German Travel Agents' Association's [DRV] "Environmental Recommendations for Touristic Destinations").

The Council of Europe pointed out back in 1973 that the coastal ecosystems must be better protected. International co-operation has thus far limited itself to action against ocean pollution in certain areas. However, particularly at the European level there have been many conferences at which the urgency of protective steps for coastal landscapes was discussed (cf. EUCC 1992, p. 10). Binding regulations are most likely to be found at the legislative level of individual countries (cf. chapter 10.2). All but five of the European states have adopted legally binding regulations for the protection of their coasts. These countries regulate coastal protection in part by way of planning systems (e.g., the Netherlands, Great Britain). The legislative framework of the various countries is applied in varying degrees. And the current status of implementation of the legal regulations, recommendations etc. also varies widely from country to country.

10.1 International Programmes and Conventions for Protecting Europe's Coastal and Marine Ecosystems

A binding convention for the countries of Europe specifically regulating the field of problems involved in "Tourism and Coastal Protection" does not yet exist. Thus far there are only agreements containing statements about sustainable tourism which are still in the process of being worked out and have yet to be negotiated in full and adopted (cf. Section D). Existing conventions and programmes are concerned primarily with ocean pollution in certain areas. The following presentation discusses the conventions and programmes for the European seas which have a direct relevance to the protection of coastal ecosystems.

10.1.1 Mediterranean

As early as 1974 the *Mediterranean region* was recognised by the *UNEP Regional Seas Programme* as a region of major environmental stress. The bordering states were to be supported in the drawing up and implementation of action plans. Under the auspices of the UNEP, the *Barcelona Convention* for the protection of the Mediterranean from environmental pollution was also adopted. The *Mediterranean Action Plan (MAP)*, a plan of action for the Mediterranean region, was decreed in 1975 on the basis of the Barcelona Convention. It is divided into *three action phases:*

- Phase I – 1975–80: resulting in the long-term *Monitoring Programme MEDPOL* (Mediterranean Pollution Monitoring and Research Programme).
- Phase II – from 1981: *concretisation and expansion* of the Monitoring Programme to four different levels (emission sources, near-coastal areas, coastal areas, transporting of emissions from the atmosphere).
- Phase III – from 1983: implementation of the MEDPOL Monitoring Programme by drawing up *national monitoring programmes.* 16 European countries are currently participating and providing data (cf. EEA 1995).

The *Blue Plan* for the Mediterranean region is part of the environmental-management component of the Mediterranean Action Plan. The Blue Plan was commissioned in 1979 and is designed to lend substantive support to the Mediterranean countries in their practical decisions regarding coastal and marine protection. The report, which attempts to describe the future of the Mediterranean countries by means of scenarios, was first published in 1989 by the UNEP, entitled "Futures for the Mediterranean Basin".

10.1.2 Black Sea and Sea of Azov

For the *Black Sea and Sea of Azov* there were, in contrast to the Mediterranean, until only recently no national or supranational regulations. It was not until 1992 that a *Convention for the Protection of the Black Sea* based on the Barcelona Convention was decreed; it was signed by Bulgaria, Georgia, Rumania, the Russian Federation, Ukraine and Turkey. The *GEF (Global Environmental Facility)*, a multilateral fund shared by the World Bank and the UNEP to finance environmental-protection steps, is designed to assist countries in drawing up action plans for implementing the Convention. In the *Odessa Declaration* of 1993, all six environment ministers of the signatory countries demanded the development of improved monitoring of emissions so as to better assess contamination and the development of co-ordinated plans for the restoration, protection and management of natural resources. In reaction to this, only recently a *project was approved by the GEF* in which short-term financial support for the analysis and evaluation of the environment, for the expansion of institutional capacities, for the drawing up of financial feasibility studies and for an *emergency-action programme* was granted. After that, emission sources are to be removed or reduced, among other things by *building sewage-treatment plants.* Other conventions for the Black Sea and Sea of Azov contain regulations on fishery (*Fishery Convention* of 1959) and water quality of rivers emptying into them (*Bucharest Declaration* of 1985) and protection and management (*Sofia Convention* of 1994).

Tourism as a cause of environmental stress thus far plays no role in the conventions on the Black Sea and Sea of Azov. In view of the huge strain on the marine ecosystems stemming from industrial discharge etc., this is only too understandable.

10.1.3 White Sea

The *White Sea* is included in the *Arctic Monitoring and Assessment Programme – AMAP* which was signed in June, 1991 by eight arctic countries, including the Russian Federation. The programme is to be implemented by an action group. The main thrust of the work will be concentrated on the study of persistent organic materials, selected heavy metals and radionuclides.

The aspect of a tourism compatible to the environment has been included in the Antarctic Agreement, but it, like other points, as well, has not yet been finalised and adopted. Here further international and national activities are needed to point out the significance of environmentally compatible touristic steps.

10.1.4 Barents Sea and Norwegian Sea

The *Barents Sea* is part of the area governed by the *Paris Convention* of 1978 (see also North Sea). Data on the Norwegian sector are sent to the Paris Commission every year. As the Convention has yet to be signed by the Russian Federation, however, comparable data for the sector of the Russian Federation are not available. The Barents Sea is also *part of the AMAP,* in which both Norway and the Russian Federation are participating. The *Norwegian Sea* is mainly covered by the signatory state of Norway for the Paris Convention. Above the Arctic Circle it is likewise part of the *AMAP.*

10.1.5 Baltic Sea

In view of the severe pollution of the Baltic Sea, the first Helsinki Convention *(Convention on the Protection of the Marine Environment of the Baltic Sea Area)* was signed in 1974 by all the states bordering it at the time. This led to the establishment of the *Helsinki Commission (HELCOM)*. Following the political upheaval in Eastern Europe and in view of the current need for international protection of the marine environment, a new, revised Helsinki Convention was signed in 1992. In the same year the *Baltic Sea Joint Comprehensive Environmental Action Programme* was adopted, by virtue of the provisions of which the HELCOM 92 signatory states, along with Norway, the Czech Republic, the Slovakian Republic, Ukraine and international financial institutions are co-operating to clean up the major sources of stress in the Baltic region. The programme, which is planned to last 20 years, will require an investment of 18 thousand million ECUs.

Since 1979 the *Baltic Monitoring Programme (BMP)* has been co-ordinating the monitoring of all states bordering on the Baltic. After two assessments conducted in the 80's, a third one concerning the environmental status of the Baltic Sea is in preparation.

In 1988 the environment ministers of the states bordering on the Baltic adopted a declaration on the reduction of the discharge of toxic or persistent organic substances, heavy metals and nutrients by 50%.

The programmes and agreements on the protection of the Baltic which already exist and the discussions and exchange processes which have been taking place among the bordering states are ideal preconditions for an expansion of the programmes by specific coastal-protection steps and for including tourism in the group of causes for the environmental stress on the Baltic. The already existing range of instruments should be utilised and expanded in implementing environmentally compatible tourism projects.

10.1.6 Northeast Atlantic and North Sea

The Oslo Convention (*Convention for the Prevention of Marine Pollution by Dumping from Ships and Aircraft*, 1972) and Paris (*Convention for the Prevention of Marine Pollution from Landbased Sources*, 1974) form the regulatory framework for protecting the Northeast Atlantic (including the North Sea) from pollution. The Oslo Convention is intended to prevent pollution from waste discharged from ships and airplanes. The agreement was ratified by Belgium, Denmark, Germany, Finland, France, Iceland, Ireland, the Netherlands, Norway, Portugal, Spain, Sweden and the United Kingdom of Great Britain and Northern Ireland and entered into force in 1974. With the exception of Finland, the states mentioned above are also parties to the Paris Convention, which was also signed by Luxemburg and the European Economic Community and has been in force since 1978. The Commissions of Oslo and Paris administer programmes and measures on their own or in co-operation to assess the ecological state of the sea or to serve to prevent, remove or lessen the pollution of coastal waters and the open sea. The effectiveness of the steps is examined on the basis of implementation reports and monitoring data. Steps are adapted to new situations where necessary.

In 1992 Belgium, Denmark, Germany, Finland, France, Ireland, Iceland, Luxemburg, the Netherlands, Portugal, the United Kingdom of Great Britain and Northern Ireland, Spain, Sweden, Switzerland and the Commission of the European Communities signed the *Convention for the Protection of the North-East Atlantic; "OSPAR Convention").* After entering into force, this framework agreement will replace the agreements of Oslo and Paris.

At the political (ministerial) level *Conferences on the Protection of the North Sea* have taken place at irregular intervals (1984, 1987, 1990, 1995) and, following the *Joint Declaration on the Protection of the Wadden Sea* of 1982, regular *Trilateral Governmental Conferences on the Protection of the Wadden Sea* have taken place (seven thus far). The resolutions of these conferences are generally adopted in the form of declarations and either implemented directly or in the form of steps internationally agreed upon by the participatory states.

All states bordering on the Northeast Atlantic which are members of the European Union must also implement pertinent *Directives of the European Communities* which directly or indirectly serve to protect the marine environment.

The regulatory framework and bodies mentioned above directly or indirectly cover programmes and steps for preventing ocean pollution by ships, the discharge of waste, offshore rigs and discharge from the land, steps for species protection and the monitoring and assessment of the state of the marine environment.

Just as described for the Baltic Sea, the existing range of instruments of the states bordering on the Northeast Atlantic and North Sea should be utilised for the concrete implementation and further development of sustainable-tourism projects. Monitoring and the existing regulatory framework should be examined to this end and further expanded.

For the *Irish Sea* the governments of Great Britain and Ireland *(Irish Sea Study Group ISSG)* have also conducted *monitoring.* In 1990 a *status report* was drawn up. Here, too, the *Oslo Convention* and the *Paris Convention* were followed (see also North Sea) as well as *EU directives* regulating the discharge of hazardous substances in surface water and the deterioration of water quality caused by it.

10.1.7 International Agreements

Of international significance for the habitats of water and wading birds is the *RAMSAR* Convention, which regulates the protection and preservation of wetlands. For "migratory species" such as whales and seals, the *Bonn Convention is a very important regulatory instrument*, as is the *Bern Convention* for wild animals and plants. This last is in turn the major basis for the implementation of the provisions of the Agreement on Biological Diversity for the Protection of Species in Their Habitats (more on this in Section D).

Of crucial significance for species and habitat protection within the EU is the *Fauna-Flora-Habitat Directive (FFHD)*. It principal goal is the *"preservation of biological diversity"* and "(...) *to maintain or restore a favourable state of preservation of the natural habitats and wild animal and plant species of interest to the Community*" (Article 2/2). The primary goal in the field of habitat protection is a co-ordination or improvement of nature-protection regulations of the Mediterranean countries, particularly with regard to unified regulations for species protection. For the protection of endemic species the main focus was also on Mediterranean endemites (cf. SSYMANK 1994). The focus of protection in Central Europe lies on the habitats listed in Appendix I of the Directive. This includes the protection of the *marine and coastal ecosystems* in Europe.

The implementation of the FFH Directive should be used by the various countries to decree effective regulations for protecting the coastal ecosystems from tourism. It should be stressed that tourism can endanger the preservation of biological diversity and thus requires special regulatory supervision which limits or, when necessary, precludes detrimental influences on species and habitats.

To sum up, the overview shows that in the area of monitoring the quality of bodies of water, much progress has been made and a good data base has been compiled. However, comparable monitoring programmes do not yet exist for coastal landscapes. Solely in the Mediterranean region have monitoring programmes been carried out since 1989 as part of the Mediterranean Action Plan, called the *Coastal Area Management Programmes (CAMP)*. An evaluation of the first results is not yet available (cf. chapter 10.6). Also, it remains unclear what role tourism is to play here as an impacting factor in coastal areas. The lack of an efficacy analysis of these programmes is an indication that thus far few concrete implementation steps have been carried out to protect coastal ecosystems.

10.2 Regulatory Instruments: Laws

In most European countries there are already legislative regulations governing the protection of coastal areas. According to NORDBERG (1994) there are basically two types of legislative regulations in Europe which have an influence on the protection of nature and landscapes on the coast. These will be described below (the discussion is based largely on the compilation in NORDBERG 1994, in: Council of Europe Colloquy on the Protection of Coastal Areas of the Adriatic Sea):

1. The one type of legislative regulation serves the protection of a certain *strip of coast* (defined in metres). The principal aim is the free public access to the coastline, but also the protection of nature and landscape, for example from building activity in certain zones. In Europe the width of the protected strip is usually from 100 to 300 m. Only Great Britain, Ireland, Belgium, the Netherlands and Finland have no such regulation.
2. The other type of legislative regulation is specifically designed to serve certain *habitats* of flora and fauna, especially of endangered species. These habitats generally include many important coastal habitats, which can, however, extend farther inland than is generally the case with coastal-strip protection. Apart from the national regulations there are also biotope-protection systems developed at the international level (cf. chapter 6.3), which are to be implemented by the individual countries.

Re 1: Legislative Protection of Coastal Strips. In the countries in which legislation is based on Roman law, the coastline is considered public property. Private ownership can therefore not extend to the coastal strip on land or water.

In *France* the coast was already legally defined in Colbert's time (17th century). The use of the coast and its protection are today regulated in the coastal law of 1986 (Loi relative à l'aménagement, la protection et mise en valeur du littoral). It stipulates that outside settled areas new buildings and infrastructure facilities may not be established within a 100-m-wide strip, measured inland from the high-water mark. An exception is made for construction and activities which are in conjunction with the coast, such as harbours and fishery. The protected coastal strip can be expanded to 200 m by local land-management or development plans if this is justified by high environmental sensitivity in this area and/or a great danger of erosion. The law also regulates near-coastal road construction. New roads must be built at least 2 km away from the coastline.

Spain regulates the use of the coastal strip similarly to France. The Spanish constitution declares that state property consists of all possessions which the law declares to be such; in any case coastal waters, beaches, the open sea belonging to the land and all the natural resources contained therein are included among them. The Law of the Coasts (Ley de costas) of 1988 defines various protected areas along the coast. First of all, there is a strip of coast 6 m wide on which the public right of passage is guaranteed; it can be expanded to 20 m. Also, outside of settled areas there is a protective zone 100 m wide extending inland from the high-water or high-tide mark, including dunes and beaches. The state government can expand this protective zone to 200 m if the regional government and communities consent. Most uses and developments are prohibited within this zone. Certain facilities which cannot be established elsewhere can be approved by the central government. The law also regulates the public access to the coast and sea from the interior. All development and building plans must include pedestrian access to the sea (every 200 m) as well as motor-vehicle access (every 500 m). A distance of 500 m inland from the coastline is considered to be a "near-coastal zone of influence" and must be borne in mind in the stipulation of protected zones.

In *Portugal* the width of the protected strip of coast is defined by law as depending on the type of coast. The regulation is complicated: the coast is public property. Outside of public settlements the protected strip is 200 m wide, starting from the cliffs and rock; in the area of dunes the protected strip is also 200 m wide, beginning at the end of the dune. On other coasts, with pebble beaches, for example, mudflat or marsh areas and in dry grassland, the coastal strip can be broadened to approx. 500 m.

In *Italy*, too, the coast is public property. The law stipulates that any building activity or other change of the natural environment is prohibited within a strip 300 m wide, measured from the high-water mark. This prohibited zone stays in effect until the regional government has drawn up a landscape plan for the region regulating the protection of the coastal strip (which stipulations would be contained in these landscape plans is not known – *authors' remark).*

Greece prohibits new building or other developments within a strip 150 m wide, measured from the high-water mark.

In *Turkey* the constitution states that all coasts are subject to the authority and legislation of the state. The law of the coasts of 1990, in its revised version of 1992, protects a strip of 100 m, measured from the edge of the coast. This means in practise that the protected strip in the area of cliffs and rocky coasts extends 100 m inward from the coastline; in areas with dunes, pebble or stony beaches, mudflats, marshes and similar areas which came about through water movement, the protective strip begins inward from behind these zones.

In *Germany* there is no general coastal protection stipulation in federal legislation. At the state level, however, such regulations can be adopted. Mecklenburg-Vorpommern, for example, protects a strip 200 m wide; in contrast, Schleswig-Holstein (Schleswig-Holstein Law on Landscape Management of 1985, article 40) has only a strip 50 m wide inland from the high-water mark. In Lower Saxony the greatest portion of the coast is protected by virtue of the protected status of the Wadden Sea National Park.

As regards nature conservation, Denmark currently has the most progressive coastal-protection legislation in Europe (cf. NORDBERG 1994). The Law on the Protection of Nature adopted in 1992 and revised in 1994 extends the protected coastal strip from 100 m to 300 m. This distance is measured from the point where permanent vegetation begins. Only within settled areas and areas which are stipulated in development plans as summer-home zones does the width of the protected strip remain 100 m. A commission is entrusted with the task of conducting a study of all coasts to ascertain where the protective strip is less than 300 m due to existing buildings and can remain thus. The city-planning law of 1993 was revised at the same time as that governing the protection of nature. In this law it is stated that "undeveloped" coastal segments are to remain a natural and landscape resource for the country. The law stipulates a coastal zone 3 km wide, within which development planning must be guided by the regulations described above. All regional and local authorities are charged to re-examine the plans they have already worked out to include this aspect.

In *Sweden* the coast is protected 100 m inland and seaward. This zone can be extended to 300 m by the provincial government. Over 70% of the Swedish coast is designated an "area of national interest". In such coastal areas the protective strip has in most places been extended to 200–300 m.

In *Norway* the coast is protected 100 m inward from the coastline.

Several *states of Eastern Europe* have also passed laws in recent years to protect their often unspoiled nature from commercial exploitation. Latvia has gone to the greatest lengths in this regard. There a strip 300 m wide extending inland and seaward from the coastline is protected from changes of any kind.

At present, according to NORDBERG (1994), there is hardly any prospect for co-ordinating coastal protection throughout Europe, e.g., the width of the coastal strips or the general protection of certain biotopes, as natural and political conditions are too different.

"One attainable objective, however, would be the *establishment of minimum coastal-protection standards determined by law* – recommendations are not enough. Since the beginning of 1992 the EU has been working on this topic, but a commonly adopted law of this kind is thus far not a prospect (...). The Council of Europe could begin an initiative and work on a set of legally binding coastal-protection instruments" (cf. NORDBERG 1994).

Re 2: General Protection of Coastal Habitats. A good example for this type of legislation is Germany. The Federal Law on Nature Protection of 1987 contains a list of generally protected biotopes. The states are obligated to lend this habitat protection concrete form in their state legislation. They are entitled to put other biotopes on the list. The states of Mecklenburg-Vorpommern and Schleswig-Holstein protect (with small variations) the following biotopes located on or near coasts, for example: dunes and sandbanks, cliffs and steep coasts, most heaths, salt meadows, salt marshes and mudflats, reed areas, dry-grass areas, bog forests, natural unregulated brooks including the area along their banks, river sources and standing bodies of water. Furthermore, Mecklenburg-Vorpommern protects its shallows. A similar list of protected coastal biotopes is in *Denmark*.

The general protection of certain coastal biotopes has clear advantages: instead of protecting the biotopes in individual areas dispersed over the land, which often entails long, tedious bureaucratic procedures, the problem is solved in one step. It would be desirable for countries which are currently restructuring their laws and/or are working on the Fauna-Flora-Habitat Directive (FFHD) described in chapter 6.3 to follow this lead. The FFHD offers the foundation for realising general biotope protection within the framework of national laws.

10.3 Planning and Monitoring Instruments

In some European countries coastal protection is mainly regulated through planning systems. In Great Britain, the Netherlands and Finland, for example, this is done using a corresponding development plan (DP). In Turkey and Greece touristic master plans are being worked out containing framework regulations for ecological and landscape planning. In Germany it is landscape planning, in particular, by means of which attempts are being made to achieve sustainable touristic development plans. The legal basis is the Federal Nature-Protection Law, described in chapter 6.2, and the special regulations of the separate states.

Various possible planning and monitoring instruments will be described below with whose aid tourism can be made more sustainable. A complete comparison of the specific planning systems of each of the countries cannot be given here, as it would exceed the scope of this project. Examples of regional, development and landscape planning in the European countries will be presented as well as international planning instruments, such as the environmental-impact assessment (EIA) and the eco-audit of a commercial operation, both of which are relevant for the sustainability of touristic projects. Then follows a brief description of geographic information systems (GIS), which are being used throughout Europe for the presentation of data relevant to planning.

10.3.1 Tourism and Regional Planning as Practised in Germany

Regional and development planning in the individual European countries varies widely in effectiveness for coastal management, as NORDBERG (1994) has shown. Detailed comparative studies are not available. Therefore only a description of the possible applications of regional planning for touristic projects will follow here, citing the example of the German planning system (based on PROGNOS 1994). According to statements by the environment ministers of the German states, it is in any case necessary "to exert an active influence on the development of a sustainable, environmentally compatible tourism. The *range of instruments of regional planning* is to be used for the implementation of the plans" (BMU 1995, p. 352). Above all, the ministers stress the significance of the *development of common guidelines for regional planning* for the coastal regions of the Baltic region. "Delicate coastal areas should not be opened up to tourism" (op. cit.).

To minimise conflicts between touristic development and the concerns of nature and environmental protection and to make full use of the recreational potential of a landscape, the instruments of landscape planning are to be utilised where available. PROGNOS (1994) describes the possibilities for action to be taken by planning using the example of the Baltic region of Mecklenburg-Vorpommern as follows:

- speedily placing extended ecosystems *under protection;*
- adapting tourism to the landscape's sensitivity and protection requirements *(zoning);*
- *development of complementary touristic functions* for coast and interior (which means that the *hinterland* is to assume increasingly "complementary" touristic functions for the coastal region; this is also a touristic development strategy in Spain: Extremadura – *authors' remark);*
- *large-scale touristic projects* are to be concentrated in a few ecologically resilient sites and as far as possible in connection with existing populated areas;
- for all plans of greater magnitude *regional-planning procedures with integrated environmental-impact assessment* are to be carried out;
- *information and awareness programmes* for population and tourists etc.

Another instrument for testing the regional compatibility of touristic projects in Germany is the regional-planning procedure (RPP). This examines the need and compatibility of a project with the conditions of the site and its environment. If severe environmental impacts are expected to be caused by the project, special environmental-impact assessments (EIS or EIA) are also to be integrated in the RPP. Touristic-development steps, for example, are routinely subject to an RPP, as are plans for motor-sport and sport-aviation facilities, large-scale camping grounds, hotels, holiday villages, pleasure and sailboat harbours etc. (cf. SCHARPF 1989).

10.3.2 Coastal Protection Through Development Planning

In the *Netherlands* and in *Great Britain* there is no legislation concerning the general protection of a defined strip of coast. This lack is in part compensated for by an effective system of development planning. In Great Britain, for example, all local authorities must draw up a structure plan for their area and in it pay heed to national policy guidelines on development planning. Among these, for instance, are guidelines on coastal planning which were adopted in 1992 by the government. The guidelines state that currently undeveloped coasts must not be used if this use can be accommodated in the interior or at already developed sites, as well. The protection plans are in turn the basis for the issuing of building permits. These can be cancelled outright. The result has been that where the structure plan dictates that no commercial or building activity may take place, natural coastal areas are hardly more valuable than agricultural land and can be bought up at a low price and then used for nature protection.

Another example where coastal protection is done by means of development planning, but with little success, is *Finland*. Here the legal regulations are completely different from those in Great Britain. In Finland the general policy is that landowners can build without a permit as long as this is occasional building of 4 to 5 houses per kilometre of coastline. Where a coastal plan exists, 10 to 12 houses may be built per kilometre of coastline. The community can, but need not, draw up a development plan for the coastal area in question. The planner can try to group the development rights of landowners differently to keep certain coastal areas free. As the coastal areas in Finland are often divided up into tiny plots, however, this is virtually impossible. If the state or the community would like to protect a natural coast from destruction, it must buy land at market prices, which are determined by the value of speculative development rights.

Most European countries have no effective planning system which includes coastal protection. On the contrary, there seems to be a strong tendency to transfer decisive authority regarding regional planning from the central to regional administrations. In view of the generally high economic value of coastal areas, this is a fundamental danger for the protection of coastal ecosystems. As the financial budgets of regional and local administrations are usually tight, it is doubtful whether local authorities wll be able to resist applications for construction sites.

10.3.3 Tourism and Landscape Planning

To minimise conflicts between touristic development and the concerns of nature and environmental protection, instruments of landscape planning can be used, if there are any in that country.

In the landscape plans in the German planning package (community landscape plans and landscape framework plans at the regional level), the ecological framework conditions for touristic development are compiled and presented in comprehensive form. An example of an important instrument of environmental precaution and security at the community and regional levels is the compatibility of natural life bases with the demands of touristic use. The landscape plan contains an inventory of the total ecological situation in the planning area and an assessment of the impairments of the natural balance to be expected from planned projects. Building from this, development ideas for the sustainable securing of the functioning of the natural balance are drawn up and presented (cf. BTE/WIRZ in: BfN 1994).

10.3.4 Tourism and Environmental-Impact Assessment (EIA)

According to the 1985 Directive of the European Commission, a number of touristic projects must be subjected to an EIA: holiday and leisure-time parks (e.g., Center Parcs and Euro Disney in France), swimming and play pools, hotel complexes, ski lifts, marinas etc. The EIA is designed to be an instrument of preventive environmental protection which appraises and evaluates various alternatives and then makes recommendations.

The environmental-impact assessment of touristic projects has proven to be useful for comparing the economic benefit of the tourism projects and the predicted adverse impacts on the environmental situation and the social and cultural conditions at the site (cf. WWF UK 1992). Possible conflicts of interest (e.g., between accommodations enterprises, restaurants/bars, residents, tour operators, environmental initiatives etc.) can be reduced or even avoided by means of an EIA by making all the significant framework conditions and facts transparent and describing them. Following the examples of Australia and New Zealand, environmental-impact assessments are already being employed in European CAMPs, as shown in chapter 10.3.

Considering that there are no standardised EIA criteria to date, the Environmental-Protection Foundation in the Netherlands (Stichting Milieu Educatie), with the support of the EU, is attempting to develop a quick, quantitative procedure for assessing touristic projects. This instrument, called the *Rapid Assessment Matrix (RAM)*, is being tested in two experimental projects in the Wadden Sea region. Final results have yet to be reached. But it is clear already that comprehensive data must be compiled to apply this method (cf. TAVERNE 1995).

10.3.5 Tourism Planning and Geographic-Information System (GIS)

By now satellite data, aided by digital image processing, are being used worldwide as an information source for statistical analysis of various environmental criteria. Further processing of these data at the digital level is an additional possible step, according to SCHÜMER (1993). Particularly suited to this task are geographic-information systems. As in environmental-impact assessments of touristic development projects and coastal-management programmes, a wealth of data is also compiled, the GIS can be very helpful to confirm scenarios, as the data are compiled in an organised and user-friendly way. The various information gathered in an EIA, for instance, can be interrelated using a GIS. In a Kenya project of the UNEP the status of the coasts is being assessed using a GIS.

10.3.6 Eco-Audit

The Commission of the European Community took up the instrument of the eco-audit in 1992 in a proposal to the Council of the Community concerning the Europe-wide regulation of systems of ecological monitoring of operations (92/C, 76/02). With the Directive of the Council of the Community adopted in 1993 (Directive (EEC) No. 1836/93) the monitoring and assessment of environmental-protection activities have become standard procedure in voluntary company environmental management.

Company eco-audits serve the monitoring of the organisation of the company's environmental protection using the documented steps taken by environmental management. According to the provisions of the EC directive, company environmental management should

- determine on the basis of self-defined environmental goals the steps, procedures and guidelines with which the environmental impacts of the activities, products, processes and potential accidents can be assessed and minimised or avoided,
- inform the public about potential environmental impacts,
- promote the goal of environmental protection with staff, customers and contract partners.

In contrast to environmental-impact assessments, with which the environmental stress of certain planning and projects is appraised in advance, eco-audits monitor the environmental impact during or after operations as well as the efficacy of countermeasures. Of major, fundamental concern to audits is constant improvement: the results of the audit of a company should always be better than the previous one. According to the principle of effectiveness, the weak points of company environmental precautions are thus removed. Eco-audits are conducted by a neutral, governmentally certified examiner, are intended to support the public-relations work of the company and, as an early-warning system, recognise critical situations beforehand and thus minimise liability risks. The preconditions for conducting company eco-audits are:

- the existence of environmental management as a separate entity,
- the formulation of company-specific environmental objectives,
- the constant monitoring and documentation of environmental impacts and protection measures.

The aim of the Eco-Audit Directive is, in keeping with the objectives of the 5th EC Environmental Programme for Lasting and Sustainable Development, to integrate the notion of environmental protection in the business world and to stress the responsibility of companies for protecting the environment. In it it is hoped that thanks to pressure from customers or consumers demanding official certification of the quality of the company's environmental protection, a "voluntary coercion" to participate will be exerted.

Carrying out *pilot projects* should help to locate existing problems in implementation and, with the pool of experience thus gained, draw up unified standards of compliance for the companies – at least in individual sectors.

The *monitoring of intra-company structures of holiday and hotel facilities* could and should be an additional sphere of the eco-audit. First examples have been tried (cf. DOMSCH 1995). Although most of the big tour operators have hired "environment managers" by this time, eco-audits have thus far not been conducted in these companies.

In 1994, the German Tourism Association (DFV) commissioned the Bureau for Tourism and Recreation Planning (BTE) to draw up a *community eco-audit* whose clear criteria could be used by the host communities as an orientation guide. The principles set forth therein were then to be put to the test in a nationwide competition, "Environmentally Friendly Holiday Communities", which took place in 1995. Just how effective the set of instruments in the community eco-audit is or can be, has not been subjected to final analysis to date (cf. BTE 1994).

On the use of the eco-audit and its further *development for touristic fields of activity* in other European countries, no further studies are known.

10.4 Economic Instruments

Apart from regulatory and planning instruments there are also approaches for solving the problems of conflicting uses on the coasts by means of economic instruments. Among these is, for example, the very successful French programme Conservatoire du Littoral, begun in 1975, in which coastal ecosystems are protected from touristic and other encroachment by purchasing the land. Thus far 37,000 hectares of natural land have been secured by such purchases. That is a good 7% of the entire French coast (cf. Example: France – Côte d'Azur and EUCC 1992).

In *England* and *Wales* some 44% of the coast are protected in various types of protected areas. In addition, long coastal segments are protected by legal nature-protection regulations. The British National Trust and similar state organisations are constantly buying up more land on the coasts in order to protect it; thus far this has come to 800 km of the coast.

At the conference of the EU ministers of the environment in May, 1994 in Santorin, economic and tax instruments for protecting nature from the influences of mass tourism were brought up and discussed. At this juncture it became conceivable that a "nature tax" could be introduced to raise funds for nature protection (MÄRKISCHE ALLGEMEINE ZEITUNG, July 25, 1994).

In the German state of Baden-Württemberg a nature-tax plan is being tested for the first time in a tourist community. Per overnight stay the visitors are required to pay 30 pfennig more. With the additional money landscape-management steps in the region will be financed (SÜDWEST-UMSCHAU, April 19, 1995, p. 18).

10.5 Informal Instruments

10.5.1 Information and Public-Relations Work

An important instrument for the implementation of plans for sustainable tourism is *information and public-relations work*. An awareness-building programme aimed specifically at tourists and the local residents promotes knowledge and understanding for sustainable behaviour and bolsters acceptance and identification with sustainable activities in the regions. Sustainable behaviour of tourists and natives is not a matter of course even in protected areas, as the Federation of Nature and National Parks of Europe (FNNPE) made abundantly clear in its 1993 report, "Loving Them to Death?" The information must be precisely targeted, and this includes environmental education in the broadest sense of the term: imparting knowledge of ecological complexes in nature and the landscape and on the influences of anthropogenic factors, the imparting of an awareness for these phenomena and a heightened sense for nature areas and nature experiences are the main thrust of information and public-relations work. Possibilities for implementation are, for example, the setting up of information centres, holding tours and courses on nature and the environment and establishing educational paths and "experience trails". Successfully implemented information and public-relations plans have already been put into effect in many of Europe's national parks (Bavarian Forest National Park in Germany; Coto Doñana, Spain; Peak National Park, England and many others) and nature parks (Sierra Norte, Spain; Haute-Sure Nature Park, Luxemburg etc.) as well as in the parcs naturels régionaux , the regional parks of France (cf. FNNPE 1993).

10.5.2 Environmental-Quality Seals

The demand for environmentally precautionary action in tourism has been increasingly recognised by people of responsibility in tourism in recent years. The obvious need for touristic environmental-protection plans has been coupled with the *demand for defined objectives*, i.e., the formulation of criteria for assessing environmental-protection efforts. An attempt at implementing clear objectives are the numerous environmental-quality seals, primarily for the hotel and restaurant sector (e.g., the *environmental-quality seal* of the German Hotel and Restaurant Association DEHOGA 1993), but also for tour operators (e.g., the Blue Swallow of the "Travel Compatibly" group, 1990), tourist communities (e.g., the Kleinwalsertal environmental seal of 1989/90) and tourist regions (e.g., environmental seal of the Tyrol). At the European level the "Blue Flag" should above all be named, which is awarded to beach segments and pleasure-boat harbours on the basis of appropriate environmental standards, and the “Green Suitcase” of the Association of Ecological Tourism in Europe (Ö.T.E.), which is designed to honour tourist destinations, accommodations and tour operators. The association advocates environmentally compatible and socially responsible tourism in Europe (cf. RÄTH 1993). Quality seals have the objective of formulating minimum demands of environmental precaution and environmental quality. Quality seals also serve as a guide for the consumer in looking for sustainable-tourism offers. However, the multitude of quality seals which already exist in the realm of tourism alone and the slight transparency of the criteria behind them is confusing to the consumer. Quality seals are criticised on all sides. According to MÜLLER (1993) environmental-quality seals only have a chance if:

- a high degree of responsibility is assumed by the person providing touristic services himself;
- the assessment criteria are defined from the outside;
- the public is included in the assessment;
- an authority of integrity regularly monitors environmental compatibility;
- a clear transparency is achieved;
- the business ethics of those awarded with the quality seal is revealed and long-distance travel destinations, particularly in the Third World, can formulate their own demands.

For entire tourist-destination areas (towns/regions) environmental-quality seals do not appear to make much sense, as the tourist services offered there are not that homogeneous (mountain holidays, beach holidays, city travel, long-distance travel etc.) and there is little point in comparing, say, possibilities and necessities of action in rural communities with those of “touristically dense regions”. A general standardisation cannot do justice to the specific conditions of individual communities. But it should be possible to stipulate minimum demands for a small number of key factors, such as sewage and garbage disposal, water quality etc. (cf. BTE 1994).

But for smaller areas with special touristic activities such as beaches, ski slopes, golf courses etc., ecoquality seals are a good idea. Here there is the possibility of a clear presentation of environmental data, and comparability is ensured. The same is true of hotels etc., for which clear, unified environmental criteria can be established which are easy to monitor (MÜLLER 1993).

10.5.3 Competitions

In order to promote environmental awareness among the responsible parties in tourism, eco-innovation competitions could for example be held. On the one hand competitions offer the chance of reducing stress on the environment by taking corresponding steps in the realm of tourism. On the other hand the prospect of reward for environmentally responsible action and the publicity this would entail offers tourism managers a new possibility of bolstering their "marketing image".

The Commission of the EU has announced that it will be awarding a *European Prize for Tourism and Environment* for 1995. The aim of the prize is to bolster the citizens of the European economic region in their responsibility to better protect the natural and cultural environment within the scope of their economic activity in the tourism sector. The prize will go to tourism destinations which have taken exemplary steps in this direction (cf. TOURISMUS-INFO 2, 03/95).

10.5.4 Data Banks for Sustainable Tourism

"The knowledge of environmental problems is very widespread today among the groups involved in tourism, and there is a willingness to act with about 20 to 30 per cent of those offering touristic services and tourists. But most of those willing to act lack concrete, practical information" (HAMELE 1993, p. 27). For this reason the Dutch Stichting Milieu Educatie (Utrecht) and the German Study Group for Tourism (Starnberg) set themselves the long-term goal in the summer of 1988 of building up a European information service for promoting a sustainable and lasting development of tourism. The result is the partner network ECOTRANS with important national organisations from the fields of tourism, transportation, nature and environmental protection and consumer groups from various European countries, as well as the founding of an association of its own, *ECOTRANS* e.V. Data are compiled from all touristic sectors.

10.6 Coastal-Management Programmes for the Protection and Preservation of Natural Resources in Europe

Coastal-management programmes comprise the comprehensive assessment of coastal systems and the formulation of objectives in regard to the planning and management or administration of the resources found there. According to STERR (1993) this process includes traditional, cultural and historic aspects of specific coastal areas. Existing conflicts of interest and use are also described. *Coastal-management programmes are geared to organising the process of sustainable development of coastal zones.* The basis for them is the observation of the existing uses and their impact on the regional structure and natural resources. Building from there, goals and steps for resource protection will be formulated and implemented in solution strategies. These will in turn be controlled via monitoring programmes and corrected if need be. The most important task of coastal management is "the (fair) distribution of the existing, usually scarce resources among the competing groups of users while bearing sustainable development of coastal regions in mind and minimising social conflicts" (STERR 1993, p. 11). The best way of realising and executing integrated coastal-management programmes is with a national or supraregional programme. "Integration" in the sense of a total coastal-management plan means:

- integration of various tasks of the management process,
- integration of the authorities and institutions at all political levels involved in coastal management,
- integration of decisions in the public and private sectors,
- integration of the economic, technical, scientific, planning and institutional spheres,
- integration of the objectives of various business sectors (e.g., tourism),
- integration of the available management capacities, expertise, promotion funds etc. (cf. STERR 1993).

In view of the change in climate and the rise in sea level, the international community at the global level and its most important organisations (OECD, UNEP, UNDP, FAO, World Bank etc.) have all come to realise that the current and long-term problems and tasks in the coastal zones of the earth can best be surmounted with coastal-management programmes. In 1992 this was reaffirmed, first by the U.N. Climate Framework Convention and then by the U.N. Conference for Environment and Development (UNCED) in Rio. In chapter 17 of Agenda 21, the UNCED stressed the urgency of establishing and implementing coastal-management programmes to ensure the sustainable development of coasts with the aid of this instrument.

Worldwide there are over 140 programmes, projects and plans for coastal management (TRUMBIC 1994). *Integrated Coastal-Zone Management (ICZM)* has to date only been implemented in exemplary fashion in *New Zealand* and *Australia* (CHUNG 1994). In *Europe, Coastal Area Management Programmes (CAMPs)* have been carried out in the Mediterranean, within the framework of the MAP, since 1989. They are limited to certain major regions, where action is particularly urgent as determined by the *Priority Action Programm (PAP). Table 24* provides an overview of the major areas of coastal-management programmes in Europe established since 1989 and of the action taken there, which depends on the specific problems of the region in question.

Project area, period of study	Activities
Croatia: Bay of Kastela 1989–1993	environmental-impact assessment, sewage monitoring, water resource management, scenarios for environmental development, geographic information system (GIS)
Turkey: Bay of Izmir 1989–1993	integrated coastal management (ICM), compilation of a coastal-resource management plan, sewage treatment, GIS.
Syria: entire coast 1989–1992	ICM, compilation of a coastal-resource management plan, scenarios for environmental development, GIS.
Greece: Rhodes 1990–1994	ICM, touristic-capacity study, sewage framework plan, environmental/noise/energy assessment and planning, scenarios for environmental development, GIS
Egypt: Fuka-Matrouh 1943–1995	environmental situation assessment, ICM, water-resource management, planning for sustainable tourism development, GIS, assessment of erosion processes and planning of steps to be taken.
Tunesia: Sfax 1994–	integrated zoning management, programme for the populated area, planning for sustainable tourism development, emission assessment, concentrating on industry, GIS
Albania: Duvace-Vlore 1993–1995	environmental situation assessment, ICM, water-resource management, planning for sustainable tourism development, GIS, environmental impact assessment, nature-park assessment and planning, protection of historically and culturally valuable areas and monuments
Morocco, Algeria, Israel, Malta, Lebanon 1994–1997	in planning stage

Table 24. MAP Integrated Coastal Management Programme. Source: compilation based on VALLEGA 1994, altered.

10.7 The Pan-European Strategy for Landscape and Biological Diversity of Marine and Coastal Ecosystems

On the occasion of the *European Nature-Protection Year 1995* and the summoning of the third *Conference of the Ministers of the Environment in Sofia in 1995*, the Council of Europe assumed patronage, in its resolutions at the International Conference on the "Protection of European Natural Heritage" (in Maastricht in 1993), over the *"pan-European Strategy for landscape and biological diversity"*. This strategy, which had been drawn up by the European Centre for Nature Conservation *(ECNC)* in Tilburg, the Netherlands, in close co-operation with the *Eastern-Europe Programme of the IUCN* and the *Institute for European Environmental Policy (IEEP)* in Arnhem, the Netherlands, aims to achieve an innovative preservation and improvement of the biological and landscape diversity of Europe which can be practically implemented. Beginning with existing legal instruments and initiatives, proposals for concrete steps are to be made. The strategy can be looked upon as a suggested way of providing more substance for a European environmental programme to be worked out by the ministers of the environment; it goes beyond the scope of an expert contribution and constitutes a political declaration of intent by the signatory member countries (cf. WASCHER 1995).

At the core of the strategy is a *plan of action* to be carried out between 1996 and the year 2000. The central topic of the strategy is the development of a pan-European, ecological network of landscapes by means of corridors and buffer zones and the *integration of the nature-conservation goals in the various socio-economic sectors* such as, for example, agriculture, *tourism* and energy. Among the various goals of the plan of action is an *ecological network of coasts* and the integration of the criteria of landscape and biological diversity in the sectors (cf. CDPE 1995; WASCHER 1995).

For action sphere 5, coasts and marine ecosystems, the following activities are planned (cf. CDPE 1995):

- 1996–2000: development and implementation of an *ecological network* for European coastal and marine ecosystems; protection of certain valuable coastal landscapes analogous to the Natura 2000 network;
- 1996–1999: development of an *integrated coastal management* relevant to land and sea uses which analyses, assesses etc. the interactions in a joint management programme and corresponding planning;
- 1996–1997: development of an *action guideline* for environmentally compatible use of the coasts with clear recommendations and practical tips for administrative bodies, project sponsors and other groups working and living in coastal regions (similar to the Environmental Codes of Conduct for Tourism of UNEP 1995).

Major regional points in European coastal protection in 1996–2000 are (op. cit.):

- protection of the coastal heaths in Northern Europe,
- protection of the estuaries in the Atlantic, boreal and Baltic coastal regions,
- coastal dunes/marshes in the Atlantic regions,
- lagoons, deltas in the Baltic regions,
- coastal wetlands, lagoons and deltas on the Mediterranean coasts,
- islands and archipelagoes of the Mediterranean region of high biodiversity and scenic value.

Furthermore, seal and sea-turtle habitats in the eastern Mediterranean region are to be especially protected and a catalogue of steps to protect eelgrass areas is to be worked out.

Section D. The Legal Aspects

11 Legal Aspects Involved in the Research Project

The substance of the legal aspect of this study is based on the description of the major areas of conflict between the requirements of tourism and species and habitat protection as well as present conflict-solving instruments and procedures in keeping with the principle of "sustainable development", to explore the possibilities of further developing international conventions using these basic findings and to submit proposals. In its *legal aspects* the project is to explore the possibilities for a "tourism protocol" and work out proposals. The legal section will thus comprise:

- an analysis of legal precedences, both worldwide and regional;
- an analysis of the Convention on Biological Diversity;
- a proposal for possible regulations at the global level to ensure sustainable tourism.

The following chapters are the final report on the legal aspect within the scope of the project. After remarks summarising the need for regulating touristic activities (chapter 12; further details in the reports of the nature-protection section of the research project), the international agreements hitherto concluded which either deal with tourism directly or are applicable to tourism will be analysed, as will the Convention on Biological Diversity (chapter 13).

After the analysis and assessment of the international regulations reached to date, the need for further international agreements on sustainable tourism and the options available to them will be discussed (chapters 14 and 15). The main focus of this discussion will be the question whether agreements in their own right or annexes to existing agreements, such as a protocol to the Convention on Biological Diversity, should be preferred in reaching further agreements on sustainable tourism. Following this the question of the level of detail of a worldwide agreement on sustainable tourism will be touched on (chapter 16). The last chapter contains a proposed text of a tourism protocol within the framework of the Convention on Biological Diversity.

12 On the Problems of Sustainable Tourism and the Need for International Regulations

Before entering into the legal questions as such, some remarks will be made to summarise the need for regulating sustainable tourism. This will serve to make this legal section more easily comprehensible and also enable this part of the study to be read alone. The remarks are based on the findings of the factual part of the project and are thus only of a summarising nature.

Tourism is one of the fastest-growing economic activities worldwide. In the past ten years it has grown at an average rate of 5.5%; there is no change to this trend in sight. Tourism ranks with petroleum and motor vehicles among the world's leading economic sectors.

Apart from historic and cultural attractions, nature is a major factor in tourism. One of the most important functions of touristic activities is recreation, and in our increasingly urbanised society, this recreation is more and more sought in nature. Attractive and intact natural surroundings are thus the basis for most tourisic activities.

Correspondingly, tourisic activities are today concentrated on the following areas:

- coastal regions, including the coastal region itself, offshore portions of the sea and islands;
- mountain regions, particularly at high altitudes, but also landscapes at medium altitudes;
- nature areas, particularly forests, wetlands, steppes and deserts, as well as arctic and antarctic areas.

The increase in touristic activity and particularly the phenomenon of mass tourism have an impact on nature and the environment and can also lead to damage if nothing is done to control it, making further touristic activity impossible. From this it follows that sustainable tourism is not possible without protecting nature and the environment, which must be ensured by corresponding regulatory steps.

In general the threat to nature and the environment emanate from the following factors:

- recreational activities;
- the providing of infrastructure for recreational activities (paths, sports facilities, harbours, ski runs, ski lifts, cableways etc.);
- the providing of infrastructure for accommodations, restaurants/bars and other services;
- the providing of basic infrastructure for transportation, utilities and waste removal;
- "indirectly induced developments" such as regional migratory tendencies, urbanisation or altered values.

Europe continues to host nearly two thirds of global tourism, although for some time its share has been waning, on a global perspective, in favour of Africa and the Asian-Pacific region. Taken alone, European tourism still had a growth rate of 3.5% in recent years. Western Europe, with 41%, had the most tourist arrivals, followed by the southern regions (29%) and the countries of Central and Eastern Europe (18%).

In Europe, the most popular tourist destinations are primarily the coasts. In the Mediterranean alone – the most popular destination in the world – 35% of all international tourists are concentrated annually. But it is precisely the Mediterranean countries which have a high biological diversity in the quantitative sense and are therefore particularly sensitive to stress. The Mediterranean region is that of the highest conflict potential in Europe.

In the light of this state of affairs, it should be indisputable that there is a need for regulations in the tourism sector to ensure that tourism is sustainable. But it should also be beyond dispute that there is a need for international co-ordination. Like almost no other economic activity, tourism is an international matter. This holds true for both sides, supply and demand. Touristic enterprises are active regionally and worldwide; similarly, the hordes of tourists move over entire regions and increasingly at the global level. Long-distance tourism, which can take tourists to destinations anywhere in the world, appears to be becoming more and more popular.

This internationality suggests the need for an internationality of regulation in order that sustainable forms of tourism can be achieved by all concerned and in all areas affected by tourism. International co-ordination can in principle be achieved in political or legal form. If it is desired, and perhaps imperative, to act effectively and without delay, because of the state of nature and the environment, binding regulations are indispensable, and this leads to the conclusion of international agreements.

13 Existing International Regulations Dealing with or Applicable to Sustainable Tourism

13.1 Preliminary Remark on the Method of Selection of the Agreements Discussed

Herein those agreements will first be discussed and analysed dealing specifically with sustainable tourism, after which those agreements will be considered which do not deal specifically with tourism, but which can definitely address partial aspects of the problem of sustainable tourism. Fundamentally, a great many regional and global agreements on nature conservation and environmental protection can be considered, whether certain activities in the area of tourism fall under their provisions or whether certain ecosystems which are important to tourism are regulated by agreements. All agreements which are in some way relevant cannot be presented and analysed here; the main focus will be on the international nature-conservation agreements which have been concluded at the regional and global levels (the texts of the agreements are taken from BURHENNE, *International Environmental Law/Multilateral Treaties*, loose-leaf edition).

13.2 Agreements Dealing Specifically with Sustainable Tourism

Agreements dealing specifically with sustainable tourism, despite the economic significance of tourism and its high environmental relevance, are a new phenomenon. Thus far there are only agreements which are in the process of being reached; none has been negotiated to completion, signed and entered into force. Of great interest in this context is the tourism protocol within the framework of the Alps Convention which is now at the negotiation stage. The Antarctic Treaty and its Protocol on Environmental Protection are further contexts in which the creation of a specific legal instrument for sustainable tourism is under consideration.

13.2.1 The Tourism Protocol Within the Framework of the Alps Convention

The Alps Convention is a regional agreement concluded by the states bordering the Alps to protect the Alpine environment (cf. DANZ, ORTNER 1993; also LORCH et al. 1995). Signed on 7 November 1991, the Convention entered into force on March 6, 1995. It has thus far been ratified by Austria, Germany, Liechtenstein and Slovenia. France announced that it planned to ratify it in November, 1995. The Convention goes back to the resolution of the First Alps Conference of 1989 in Berchtesgaden and is a framework convention to be informed by individual provisions through special agreements called protocols. The topics to be specifically regulated in the protocols are named in Article 2 of the Convention. One of them is tourism.

At the Third International Alps Conference, on December 20, 1994, the first three protocols of the Alps Convention were signed. They concern:

- nature protection and landscape conservation;
- country planning and sustainable development;
- mountain agriculture.

The protocols were signed by Germany, France, Italy, Monaco, Slovenia and the European Community. Austria and Switzerland have thus far not signed (cf. *Umwelt* no. 2/1995, p. 57).

Draft of the Tourism Protocol. The Tourism Protocol is available as of now in draft form (excerpts of the text of the draft of 15 November 1995 are in the Appendix); the negotiations have not yet been fully completed.

In the resolution of the First Alps Conference it is stated that the development of tourism must be made compatible with the demands of nature and environmental protection (no. 58), that particularly environmentally harmful forms of tourism should be restricted or banned (no. 59), that by mutual agreement tourism-free zones are to be designated, further development in glacier areas and particularly sensitive ecosystems should not be permitted and stress from winter sports should be reduced (no. 60), that a further expansion of touristic infrastructure can only be permitted after conducting regional and environmental-impact assessments (no. 61) and that a comparative survey of the present situation must be conducted (no. 62).

The contracting parties accordingly pledge in Article 2 of the Alps Convention to take steps with the objective of restricting environmentally harmful touristic activity, to designate zones free of touristic activity and in general to reconcile touristic and leisure-time activities with ecological and social requirements. From this alone it may be deduced that currently existing environmentally harmful activities in the realm of tourism are to be restricted; it is not enough simply to forgo embarking on new environmentally harmful activities in future. This is significant in assessing the present draft of a tourism Protocol (cf. also LORCH et al. 1995, p. 48).

This draft of a tourism protocol states that it is an objective of the protocol "to contribute through sustainable tourism to a sustainable development of the Alpine Region" (Article 1). The interests of the local population and the holidaymakers should be given equal consideration (ibid.). The draft advocates increased international co-operation, particularly among the regional bodies (Article 2). The contracting parties also pledge to be mindful of the objectives of the protocol – ensuring sustainable tourism – in their other policies (Article 3).

The specific steps demanded by the contracting parties concern, inter alia:

- the regulation of the supply side by models, development programmes and sectoral plans (Article 5);
- orientation of touristic development to the concerns of nature conservation and landscape management (Article 6);
- the promotion of high-quality tourism, with special consideration given to ecological requirements (Article 7);
- regulating the streams of tourists by planning and organisational steps (Article 8);
- adjusting touristic development with the specific environmental characteristics and available resources of a region through prior assessment, if need be (Article 9);
- designation of quiet zones in which no touristic development shall take place (Article 10);
- a control of the building of ski lifts, also requiring consideration of ecology and landscape protection (Article 12);
- restriction of motorised traffic in tourist centres and the promotion of public transportation (Article 13);
- construction and maintenance of ski runs and artificial snow production, with special attention devoted to environmental aspects (Article 14);
- sports activities of tourists, which are to be regulated and if need be prohibited (motor sports outside public transportation routes, Article 15) and
- skydiving outside airports, which is to be banned "or at least severely restricted" (Article 16).

In the closing negotiations of the year 1995, the provisions on artificial snow production and the control of sports activities were most controversial, so changes in the proposed texts are still possible.

The protocol on tourism within the framework of the Alps Convention is worthy of special attention because it constitutes the first example of an international instrument concerning the entire problem of the environment and tourism and illustrates the possible factors to be regulated, albeit in a specifically Alpine context.

13.2.2 Considerations on a Legal Instrument for Sustainable Tourism Within the Framework of the Antarctic Treaty and the Protocol Concerning Environmental Protection

Within the framework of the Antarctic Treaty of 1959, a Protocol on Environmental Protection in the Antarctic was concluded in 1991 (excerpts in the Appendix). It will enter into force after all the consultative states of Antarctica (which now number 26) have ratified it. The Federal Republic of Germany created the parliamentary preconditions for its ratification in the summer of 1994 and decreed the law of approval and the law of execution (cf. *Umwelt* no. 10/94, p. 365).

Beneath the umbrella of the Antarctic Treaty and the protocol, a particular legal instrument for regulating tourism in the Antarctic is being discussed. Thus far a legal instrument of this kind has not been drafted, also because opinions differ as to whether such an independent international instrument is needed. Some consultative states apparently think that the Antarctic Treaty and the protocol of 1991 were sufficient at the international level and that further steps would have to be taken at the national level.

A regulation on tourism proper could be a further annex to the environment protocol. It would also be conceivable to draw up a tourism protocol which complements the Antarctic Treaty and the environment protocol and would thus accompany the two instruments. But then it would have to be ensured in a suitable way that the rules of the treaty and the environment protocol, which already exist, be heeded. But in this regard further developments must still be awaited.

The Protocol on Environmental Protection pursues the objective of comprehensively protecting the Antarctic's environment and the dependent and interdependent ecosystems and preserving the Antarctic as a "nature reserve".

A number of principles for the planning and execution of activities in the Antarctic are stipulated, among them the obligation to carry out environmental-impact assessments (cf. Article 3, section 2 c). The activities under scrutiny include tourism (cf. Article 3, section 4 and Article 8, section 2).

Details are set forth in the annexes to the protocol. The annexes are concerned with:

Annex I	EIA,
Annex II	preservation of Antarctic fauna and flora,
Annex III	removal and processing of waste,
Annex IV	prevention of ocean pollution,
Annex V	protection and administration of areas (protected areas).

The Treaty and environment protocol already regulate touristic activity in the Antarctic as follows:

- protecting the environment shall be a "crucial consideration" in the planning and execution;
- impairments to the environment and ecosystems shall be "limited";
- a number of particularly adverse impacts on the environment and ecosystems of the Antarctic shall be avoided (cf. the list in Article 3, section 2 b);
- activities in the Antarctic shall first be subject to an EIA which conforms with Annex I;
- in the planning and execution of touristic activities the regulations and prohibitions in Annex II concerning the preservation of flora and fauna shall be respected;
- similarly, the regulations on the removal and processing of waste in Annex III shall be respected;
- adherence to the special stipulations in Annex IV regarding the protection of the marine environment shall be ensured;
- finally, restrictions may result from the regulations concerning particularly protected and particularly administered areas, which must be respected by touristic planning.

At the meeting of the consultative parties in 1994 in Kyoto, it was agreed to make recommendations for sustainable tourism. In recommendation XVIII-1 on tourism and non-governmental activities, the increase in touristic activity in the Antarctic is stressed, as is the fact that "practical guidelines" are needed so that travel to the Antarctic can be planned and carried out in the best possible way. The representatives of the consultative states recommend their governments to distribute the "Guidelines for Those Organising and Carrying Out Tourism and Non-governmental Activity in the Antarctic" as widely and quickly as possible and to urge the groups in question to act in accordance with the guidelines.

The "Guidelines for Visitors in the Antarctic", which are intended to ensure that the provisions of the Antarctic Treaty and the protocol are adhered to, contain the following imperatives (cf. also the excerpts in the Appendix):

- Protect the wild species in Antarctica!
- Respect protected areas!
- Respect scientific research!
- Be aware of safety precautions!
- Preserve the untouched state of Antarctica!

The Guideline explains in detail how these imperatives are to be implemented.

The guideline for tour operators informs on the most important obligations of travel organisers and ship owners as well as on the procedures to be adhered to by both groups. These are concerned with travel planning, the stay in the area governed by the Antarctic Treaty and the termination of the journey. The goal of the guideline is ultimately to establish the principles, procedures and obligations to protect the Antarctic environment as far as possible for the increasingly significant segment of Antarctic tourism even before the environment protocol enters into force.

After the passage of the law to implement the environment protocol, German law now dictates that all activities in Antarctica, including touristic activities, must obtain government permission. Permission is issued by the Federal Agency of the Environment. Depending on the expected environmental impact, either an environmental-relevance assessment or an environmental-impact assessment must be conducted. The law of implementation also contains substantive regulations on the protection of the environment contained in the protocol, provisions and prohibitions, things which must be respected and not removed etc.

13.3 International Conservation Agreements Applicable to Tourism

13.3.1 Regional Agreements

Europe. In Europe there is by now a number of international agreements on nature protection. Some of them are in force over vast regions (all of Europe), some of them are applicable to certain areas (the Mediterranean, North Sea, Baltic, Alps). There is also overlapping, such that pan-European agreements are in force alongside special agreements which only concern one certain area. Pan-European agreements are, for example, the Bern Convention or, albeit not in force in the exact same area, the directives of the EU on nature conservation (bird protection, flora-fauna habitat). In their respective areas of effectiveness, it can happen that other agreements also take effect at the same time, such as the Mediterranean agreements or the Alps Convention and its protocols (once they enter into force) or the agreements on the Baltic (cf. in detail DE KLEMM 1993).

Europe	■ Bern Convention, 1979 ■ Barcelona Convention / Protocol on Protection Areas, 1986 ■ Convention on the Northeast Atlantic, 1992 ■ Baltic Sea Convention, 1992 ■ Black Sea Convention, 1992 ■ Alps Convention / Protocol for Nature Conservation, 1994
Caribbean	■ Caribbean Convention on the Protection of the Wider Caribbean Region, 1990
East Africa	■ East Africa Convention / Protocol on Protected Areas, 1985
South Asia	■ ASEAN Agreement on the Conservation of Nature, 1985
South Pacific	■ Convention on Conservation of Nature in the South Pacific (Apia Convention), 1976 ■ Convention for the Protection of the South Pacific Region (Noumea Convention), 1986

Table 25. Nature-protection agreements relevant to tourism / regional (selection).

Europe as a Whole: Bern Convention. The Bern Convention, a convention of the Council of Europe, was signed in 1979 and entered into force on June 1, 1982. It is concerned with preserving the wild plant and animal species living in Europe and their natural habitats. Like any measure to protect nature, this convention also has a cross-sectional effect: it includes a number of activities which impair or could impair the things the Convention is designed to protect. Among these activities is tourism, be it the activities of the tourists themselves or other activities in connection with tourism. In all activities the concerns of the protection of plants and animals and their habitats are to be respected, as set forth in the Convention. Article 3, section 2 obligates the contracting parties to respect the preservation of wild plants and animals in their planning and development policies. Strictly protected plant and animal species require special consideration (Article 4). The contracting parties pledge to establish and implement removal regulations (Article 5 for plants, Article 6 for animals). Animal species which can be utilised should be done so in a way which does not endanger their survival (articles 7–9).

Thus, the Convention requires the member states to issue regulations which include tourism, even if it is not expressly mentioned.

European Union. In the area of the European Union there are also legal instruments for nature conservation which are relevant to tourism. These are the directives decreed on the basis of the EC Treaty, no. 79/409/EEC of April 2, 1979 on the preservation of wild bird species and no. 92/43/EEC of May 21, 1992 on the preservation of natural habitats and wild animals and plants (the Fauna-Flora-Habitat Directive, or FFH Directive for short; texts of both directives are in STORM/LOHSE, EG-Umweltrecht).

The objective of the directive on the preservation of wild bird species is, apart from management, regulation and utilisation, the protection of these species (Article 1), in particular by preserving or re-instating a sufficient diversity and a sufficient magnitude of the habitats (Article 3, section 1). The member states are in particular obligated to preserve or re-instate the habitat locales and areas, to care for habitats in and outside of protected areas and ensure that they are ecologically sound, to restore destroyed habitats and create new ones (Article 3, section 2). For species under special protection (which are named in Annex I of the directive), special steps are required, which includes protection of their habitats, as well (Article 4).

With the protection of habitats, the member states also assume the obligation to regulate and monitor activities which can lead to an impairment of habitats. Thus, touristic activities are in principle one such activity, so that here, too, an indirect application to tourism results.

The Fauna-Flora-Habitat Directive is geared to preserving the natural habitats and wild animals and plants and is thus an important instrument for protecting biodiversity in the area of the member states of the European Union.

The Directive intends to reach this goal by establishing a European network of special protected areas ("Natura 2000"). In these protected areas the member states are to take all necessary steps to preserve these habitats; for any plans and projects environmental-impact assessments must first be conducted (Article 6). By requiring this "plan EIA", the Directive is generally believed to have gone even farther than the directive on environmental-impact assessments. The steps which are being taken on the basis of this Directive should also be borne in mind in regard to touristic activities.

The Alpine Region. In the Alpine region there will not only be a specific tourism protocol under the umbrella of the Alps Convention, but also one on nature conservation and landscape management. This protocol was signed in December, 1994 by most of the countries of the Alps (except Austria and Switzerland) and by the European Union. As an instrument of cross-sectional nature, it will in turn include all activities and thus tourism, too. The draft provides for obligatory surveys of the status of nature conservation (Article 5), landscape planning (Article 6), integration of this planning in general planning (Article 7), a regulation of impacts (Article 8), basic protection (general protection of nature and landscape, outside protected areas, as well; Article 9), regulations on protected areas (articles 10–12), on species protection (Article 13), bans on removal and commerce (Article 14) etc.

The protocol is effective only in those countries which ratify it. It is entirely possible that countries will ratify the tourism protocol, in which touristic activity is the main focus, but not the nature-protection protocol, which aims to make all activities in the Alps compatible with nature. (On the assessment of the protocol within the framework of the Alps Convention cf. LORCH et al. 1995, p. 45).

Mediterranean Region. In the Mediterranean region, special regulations on environmental and nature protection exist which are also relevant to tourism.

The Barcelona Agreement, a regional agreement within the scope of the UNEP Regional Seas Programme, provides a framework for the protection of the Mediterranean. It is likewise implemented through individual protocols. The important protocol in this context is the protocol on protected areas of April 3, 1982, which entered into force on March 23, 1986. In this protocol the parties pledge to take steps to protect the natural resources and natural landscapes of the Mediterranean region and to preserve the cultural heritage of this region. The establishment of protected areas is also provided in this protocol and the obligation to take the necessary steps to preserve them. The steps set forth in this protocol are also relevant to tourism.

North Sea Region. For the North Sea, there was for a long time no comprehensive agreement on environmental protection. In the past agreements were concluded dealing with individual aspects of environmental protection in the North Sea, in particular with pollution from land-based sources or pollution of the North Sea by dumping or burning waste in it.

Since September 22, 1992 there is the Convention for the Protection of the Marine Environment of the North-East Atlantic, which also applies to the North Sea. When it enters into force it will replace the conventions of Oslo and Paris. It prohibits the burning of waste on the sea and, basically, also its discharge. It defines the precautionary principle and the polluter-pays principle, anchoring both in international law. It defines the "state of the art" and the "best environmental practise" and provides that they be applied. It enables legally binding resolutions to be adopted and allows for regionally graded steps to be taken. Moreover, the Convention can be expanded in its sphere of application by attaching annexes to it. The obligations of the Convention are of a general nature regarding all activities and thus tourism, as well. In particular, coastal tourism is included, inasmuch as impairments of the North Sea are to be feared from it. Once the Convention enters into force, it will replace that of Oslo (dumping of waste) and that of Paris (pollution from land-based sources).

For the North Sea region, the trilateral co-operation of the states bordering the Wadden Sea – Denmark, Germany and the Netherlands – is also significant. This co-operation is based on a joint declaration made in 1982 and results in decisions at the governmental level which can also be significant for touristic activities. At the 7th Trilateral Government Conference from Nov. 29 to Dec. 1, 1994, a joint trilateral co-operative region was delineated and also a joint protected area stipulated. Moreover, the basic principles for working out a joint management plan for the Wadden Sea area were defined and the goals envisaged for the various habitats of the Wadden Sea identified. In addition, joint reporting on important parts of the Wadden Sea as part of the protected-area system Natura 2000 within the scope of the Fauna-Flora-Habitat Directive as well as joint principles for the protection of seals and small whales were resolved (cf. *Umwelt* no. 1/1995, p. 13). These resolutions and conservation agreements can also have effects on tourism in these areas.

The various North Sea conferences have adopted resolutions which are relevant to tourism. At the Third International North Sea Conference of 1990 a joint declaration of the Wadden Sea states was adopted which addressed tourism. The states agree to work out guidelines for the cities to reduce the stress on the North Sea stemming from, among other factors, tourism.

Baltic Region. For the region of the Baltic Sea, the Helsinki Convention has existed since 1974 with the objective of protecting the marine environment of the Baltic region from pollution from all sources. The Convention provides that the Helsinki Commission act as its operative body.

The new Helsinki Convention signed in 1992 is a revised version of the Convention of 1974. Particularly noteworthy among its new provisions are the regulations on pollution from land-based sources, environmental-impact assessments, which are mandatory for all activities of considerable pollutive potential, on nature conservation and biodiversity. Nature conservation and protection of the biological diversity of the Baltic region are expressly named for the first time, albeit not treated in detail.

Expert decisions of the Helsinki Commission are handed down in the form of "recommendations" which are politically binding, but non-binding with regard to international law. They interpret the fundamental tenets of the Convention in more detail.

Black Sea Region. For the Black Sea a Convention was concluded on April 21, 1992 in Bucharest. It is likewise a convention to control marine pollution coming from all sources. The obligations are general, so that they are applicable to all activities having an impact on the marine environment and thus to touristic activity, as well. However, the text of the Convention contains no specific steps which the states are to initiate. This must be made concrete either by national legislation or in additional international instruments (protocols). Protocols of this kind are expressly referred to in the Convention (cf. Article V, section 3).

South Pacific Region. In the South Pacific region there are two agreements which could be relevant. First, the Convention on Conservation of Nature in the South Pacific (Apia Convention) of 1976, obligating the contracting parties to establish protected areas. The states are obliged to take corresponding protective steps, e.g., forbidding removal, regulation of visitors, utilisation, control of the introduction of alien species etc.

The second agreement is the Convention for the Protection of the Natural Resources and Environment of the South Pacific Region (Noumea Convention) of 1986. The Convention is a part of the UNEP Regional Seas Programme. It is thus a marine-protection agreement. It does not include the interior and archipelago waters of the contracting parties. The obligations are geared, as in most regional agreements of the UNEP, to reducing marine pollution coming from all sources, including land-based ones. The Convention also expressly includes marine-environmental stress stemming from tourism.

There are two protocols within the scope of this Convention, one on waste removal, another on co-operation in pollution emergencies. All three agreements are in force. The agreement is based on an Action Plan adopted in Rarotonga in 1982.

In 1993 a further agreement was concluded to establish the South Pacific Regional Environmental Programme (SPREP). The agreement establishes the SPREP as an international organisation but contains no substantive provisions on nature conservation.

The Caribbean. The Caribbean is also a part of the UNEP Regional Seas System. In 1983, once again on the basis of an Action Plan, the Convention on the Protection and Development of the Marine Environment of the Wider Carribbean Region (Cartagena de Indias) was adopted, together with a protocol on co-operation in oil accidents.

In 1990 the Protocol on Specially Protected Areas and Wildlife was also adopted. In contrast to the Convention and the first protocol, the protected-areas protocol has yet to enter into force. The protocol mentions tourism explicitly, as it demands in Article 5 the regulation of touristic and recreational activities which could threaten the ecosystems of the protected areas and the survival of endangered species.

East African Region. East Africa is likewise covered by a UNEP regional agreement. In 1985 an action plan was adopted for the region, as were the Convention for the Protection, Management and Development of the Marine and Coastal Environment of the Eastern African Region, the Protocol Concerning Protected Areas and the Protocol Concerning Combatting Marine Pollution. All of these instruments have yet to enter into force.

Of particular significance for the subject of environmentally compatible tourism is the protected-area protocol, as here once again protection obligations could result which are relevant to tourism.

South Asian Region – ASEAN States. For this region there is the ASEAN Agreement on the Conservation of Nature and Natural Resources of 1985, but it, too, has yet to enter into force.

The Agreement is a comprehensive document on the conservation of nature and natural resources. It provides for mandatory protection of plants and animals and their habitats, including activities which could have a detrimental effect on them. Here, too, touristic activities are relevant.

13.3.2 Conservation Agreements Applicable at the Global Level

Agreement on Wetlands of International Significance (Ramsar Convention). The Ramsar Convention of 1971, in force since 1975, is targeted at protecting wetlands of international significance, in particular as habitats for water birds and waders (for details cf. KOESTER 1989; IUCN 1991; DE KLEMM 1993).The contracting parties are required to designate at least one such area as being internationally significant. The wetlands, which are included on the international list, must be preserved; the other wetlands are to be "reasonably utilised" by the states. This convention also contains general implications which are fundamentally applicable to tourism.

Article 3 sets forth the main obligations of the parties. They "shall formulate and implement their planning so as to promote the conservation of the wetlands included in the List, and as far as possible the wise use of wetlands in their territory." In cases where the wetlands on the list are ecologically threatened, informing the governmental agencies and the responsible international secretariat must be ensured.

Global	Ramsar Convention on the Protection of Wetlands, 1971
	World Heritage Convention, 1972
	Washington Convention on Illegal Trade in Endangered Species, 1973
	Bonn Convention on Migratory Species, 1979
	Convention on Biological Diversity, 1992

Table 26. Nature-protection agreements relevant to tourism/global (selection).

World Heritage Convention. The UNESCO Convention concerning the Protection of the World Cultural and Natural Heritage, adopted in 1972, in force since 1975, is geared to protecting sites of outstanding natural beauty (for details cf. De KLEMM 1993). Similarly to the Ramsar Convention, it can be relevant to the problem of tourism and the environment even though no specific regulations and steps regarding tourism are named in the Convention. The protective obligations in regard to the sites to be preserved are what make the Convention relevant to tourism. The steps to be taken also impact touristic activities.

In the Convention the steps are described in very general terms, however. Article 4 states that the contracting state assumes the task of identifying, protecting and preserving the existence and value of the heritage located in its territory and hand it down to future generations. The contracting state "will do all it can to this end", if need be with international support. Article 5 of the Convention states that a state party to it must pursue a policy geared to giving the cultural and natural heritage "a function in the life of the community and to integrate the protection of that heritage into comprehensive planning programmes." More detailed preservation steps are not named; nonetheless it is very clear that touristic activities must be assessed on the basis of these framework conditions.

Washington Convention on Trade in Endangered Species (CITES). The Washington Convention on species protection regulates one aspect of nature conservation: international trade in endangered animal and plant species. International trade is defined in the Convention as the import and export of plant and animal species by anyone at all, which thus includes tourists. In this sense, the obligations set forth in the Convention also apply to touristic activities.

The Convention obligates the parties to establish export and import inspection of varying stringency, depending on the endangered status of the species in question. The endangered species are subdivided into three categories: Annex I contains all the species threatened by extinction; Annex II all species requiring inspection; Annex III all species which are designated by one contracting party as in need of inspection, thus requiring the involvement of other contracting parties. The Convention describes the requirements and procedures for export and import inspection in relative detail. However, it should not be forgotten that international commerce is only one aspect of the problem. Therefore only one aspect of tourism is affected: the transport of specimens of endangered species across international borders.

Convention on Migratory Wild-Animal Species – Bonn Convention. The Convention on the Conservation of Migratory Species of Wild Animals of 1979 regulates mandatory protection of certain animal species. It is relevant here because this convention, too, contains mandatory steps for the protection of species and their habitats which include a number of activities with potentially adverse impacts on the animal species, tourism among them.

The Convention also makes distinctions according to the state of endangerment of the species in question. For endangered species (Appendix I), the Convention itself stipulates preservation steps: the preservation of habitats, control of activities with adverse effects on stock, elimination or at least minimising of adverse effects, preventing other negative influences from taking place etc. For species which are "in an unfavourable state of preservation" (Appendix II), the Convention provides for special agreements, for whose formulation guidelines are provided.

13.3.3 Convention on Biological Diversity

The Convention on Biological Diversity, signed in Rio in 1992 and in force since December 29, 1993 and as of today (the end of 1995) ratified by over 130 states, placed international nature-conservation efforts on a new foundation (for details cf. BURHENNE-GUILMIN 1993; BURHENNE-GUILMIN/GLOWKA 1994; GLOWKA et al. 1994). The Convention is designed to be more comprehensive than the agreements concluded hitherto regarding international nature conservation, all of which only protect certain aspects of it. With the preservation of biological diversity, a broader protection objective was chosen: here the aim is the preservation of the diversity of life on earth, which can only come about through comprehensive protection of habitats. Moreover, biological diversity is broadly defined in the Convention as genetic diversity, the diversity of species and the diversity of ecosystems.

But the Convention is more than an instrument of nature conservation. It is at the same time an agreement of great economic significance. It involves accessibility of genetic resources, the accessibility of technologies, financial transfers from the industrial to the developing countries. These factors are also reflected in the objectives set forth by the Convention (cf. Article 1).

In view of the broad objectives, the obligations are very general. Special problems are to be regulated in further agreements, just as the Convention leaves hitherto existing agreements unaffected, "except where the exercise of those rights and obligations would cause a serious damage or threat to biological diversity" (Article 22, section 1; excepted are agreements on the law of the sea, Article 22, section 2). This means that along with the Convention on Biological Diversity, the agreements on international nature conservation named hitherto are in principle still in force and applicable on a regional as well as a global basis. However, there remains the task of considering ways of co-ordinating all of these agreements with the Convention on Biological Diversity.

On the other hand, the scope of the Convention, particularly with regard to the protection of biological diversity, is considerable. It contains mandatory provisions on the identification of biological diversity, planning, environmental-impact assessment, in-situ conservation, ex-situ conservation, sustainable use, research and training and the building of public awareness – in a word, a systematic approach is established.

Many of these obligations are also relevant to touristic activity, even though tourism is not expressly mentioned:

- The identification of processes and activities which (can) have considerable adverse impacts on the conservation and sustainable use of biological diversity provided for in Article 7 of the Convention will have to be heeded by touristic activities.
- The national strategies for preserving biological diversity (Article 6) must also be mindful of those factors which constitute a potential threat to biological diversity and are thus in need of regulation.
- The environmental-impact assessments to ensure that activities and also plans, programmes and policies have no adverse effect on biological diversity (provided for in Article 14 of the Convention) are relevant to a great number of touristic activities.
- The regulation of in-situ conservation steps in and outside of protected areas (Article 8 of the Convention) also includes touristic activities, for protected areas and other nature areas important to biological diversity are highly significant to tourism, just as tourism can in turn be a critical factor for these areas.

13.4 Political Developments Which Can Be Significant for Further Law-Making

A few recent political developments will be pointed out here to show that tourism is increasingly being viewed as a phenomenon in need of more control. They could also be significant for the development of further legal instruments to achieve sustainable tourism, either at the regional or global levels.

13.4.1 European Union

Within the framework of the European Union, sustainable tourism is taking on increased significance. The 5th Action Programme of 1992, entitled

"Towards Sustainability" (for its final form cf. Commission of the EC, document COM [92] 23) devotes an entire chapter to tourism. Special emphasis is placed on tourism in coastal areas and mountain regions taking place in the form of mass tourism and which the programme expects will further increase. *"It will be necessary therefore to develop national and regional integrated management plans for coastal and mountain areas"* in the Community (p. 38).

The programme names as elements of these strategies the monitoring of real-estate development, the establishment of stringent regulations for new construction, combatting illegal building activity, management of private transportation routes, diversification of tourism, strict implementation and enforcement of environmental regulations in regard to noise, drinking water, bathing water, sewage and air pollution, the creation of buffer zones around ecologically sensitive areas such as wetlands and dunes, better staggering of summer holidays, building greater awareness and environmental education for the local populace and tourists and environmental education and training for people engaged in business in the tourist areas.

The programme concludes, "It is essential to place future growth of tourism within the framework of sustainability (...). Recent examples such as the algal plague in the Adriatic Sea, which cost an estimated 1.5 billion ECUs in lost revenue from tourism and fishing in 1990, indicate clearly that the environment constitutes a very important economic resource, requiring to be well maintained and protected" (p. 39).

The Council of Environment Ministers also took up the topic of tourism and the environment in 1994 (Informal Council of Environment Ministers in Santorini, May 13–15, 1994). The environment ministers state in their declaration that tourism constitutes a major sector for the implementation of steps to achieve sustainable development, that an integrated approach is needed to incorporate environmental considerations into touristic development plans and programmes, that the environmental impacts of touristic activities must be limited and that tourists must acquire the necessary environmental awareness.

13.4.2 Council of Europe

Within the framework of the Council of Europe "Recommendations for a General Policy for Environmentally Compatible and Environmentally Friendly Development" were drawn up in 1994 (the text of the proposed recommendations is in the Appendix). The recommendations contain a number of principles of environmentally compatible tourism, among them the need for environmental-impact assessments for touristic activities, being mindful of environmental carrying capacity, the principle not to further contribute to environmental stress through tourism or mandatory eco-audits. Steps are recommended for the implementation of these principles.

13.4.3 Charter for Sustainable Tourism

At the World Conference on Sustainable Tourism of April 27–28, 1995 in Lanzarote, a Charter for Sustainable Tourism was adopted (its text is reproduced in the Appendix). The Charter contains a total of 18 general principles of sustainable tourism which also can be considered as elements for further legal instruments in this field.

14 On the Question of the Need for Further International Regulations on Sustainable Tourism

14.1 Result After Assessing Existing or Currently Planned Regulations

The assessment of existing international law and such measures currently in preparation yields the result that there is as yet no specific agreement on sustainable tourism. At the regional level various agreements are in the process of being worked out (Alps tourism protocol; possibly a legal instrument within the scope of the Antarctic Treaty).

This does not mean that tourism is completely deregulated internationally in regard to sustainability. A number of agreements on more general topics, particularly the regional and global nature-conservation agreements, are applicable to individual aspects of touristic activity. From the nature-conservation agreements the obligation may already be inferred to protect nature reserves and endangered animal and plant species from impairments caused by touristic activity, as well. Obligations to control the behaviour of tourists and the impacts of infrastructure steps are also implied. The nature-conservation agreements also entail the obligation of assessing large-scale touristic projects as to the environmental impact they may have.

Nonetheless, the fact remains that according to international law in force today sustainable tourism is regulated only indirectly and in certain aspects, but not as a major topic and in comprehensive form.

14.2 Are Further International Regulations on Sustainable Tourism Needed?

With that the question whether new specific agreements on sustainable tourism are necessary or at least appropriate is not yet sufficiently answered. The question is not only a legal one, however, but also – and perhaps primarily – a political one. From the legal standpoint at least this much can be said:

Assuming that touristic development is in need of control as regards its environmental sustainability and that for this needed control legally binding regulations are also required, it is indisputable that international law as it exists today is insufficient. The references to it in existing agreements, which are only indirect or affect individual aspects, in no way do justice to the significance of tourism as one of the world's prime economic sectors.

On the basis of the currently available data, it may be assumed that tourism is that sector of the global economy which is on the one hand most dependent on an intact nature and on the other hand has such a strong impact on nature. There are thus good reasons for regulating this activity in specific agreements.

Global as well as regional agreements could be suitable to this end. Both are not alternatives, they are both necessary. Of particular urgency is a global agreement, first of all because of the global dimension of tourism and secondly because of the influence a global agreement could have on the development of regional agreements. At the global as well as regional levels independent agreements on sustainable tourism are just as meaningful as tourism agreements within the scope of already existing agreements. This will be treated further in chapter 15.1.

14.3 Possible Impediments to Agreements on Sustainable Tourism

The conclusion of specific agreements on sustainable tourism could be countered by various objections and impediments, which, however, need not be of a legal nature.

First, it could be argued that there is already such a great number of agreements and that further agreements would be not only ineffective, but also counterproductive. The rejoinder to this would be that there is no alternative to the creation of legal regulations for a state of affairs in need of regulation. Inasmuch as regulations exist in tentative form, they must be co-ordinated and further developed to form a meaningful whole.

A second objection could be that the existing regulations are sufficient to make touristic development sustainable. The answer to that, in keeping with what has already been stated, would have to be that this is not the case. The agreements concluded thus far on nature protection have scarcely been able to control touristic developments in recent years. The Convention on Biological Diversity does indeed contain a comprehensive programme in its own right, but without specific regulations regarding tourism it will hardly be able to make it sustainable.

A third objection could be that in tourism the states have thus far been very reluctant to introduce legally binding regulations internationally. Developments in international law in this field have thus far certainly corroborated this view, but this does not make the argument compelling. It is highly questionable whether the reluctance the states have hitherto displayed to regulating tourism internationally as to its sustainability did full justice to tourism's role as one of the leaders in the global economy or if the problems of tourism have on the contrary been underestimated worldwide.

Even if it were true that the reluctance of the states in regard to agreements on sustainable tourism were a great impediment, the idea of one or several agreements on sustainable tourism could nonetheless be further pursued. The development and formulation of such agreements could be valuable in itself. This process could lead to the problem of sustainable tourism being further and more intensively discussed. Legal pioneering work could be done, as well.

15 Options for International Regulations on Sustainable Tourism

Below the options for new international regulations on sustainable tourism will be discussed. The question will be addressed whether specific agreements on sustainable tourism are better suited than additional agreements within the scope of existing international agreements. It will also be discussed whether new regulations are recommended at the global or regional level or at both.

15.1 Agreements Dealing Specifically with Sustainable Tourism

The first possible option is that of agreements concluded specifically on sustainable tourism. This is possible at the global as well as regional levels. The necessary substantive and institutional regulations would have to be formulated in agreements designed for this specific purpose, without attaching them to other existing agreements. With that, regulatory agencies would come into being which would be responsible for sustainable tourism. A number of possibilities are conceivable for structuring them; individual aspects can be regulated or the problem of sustainable tourism as a whole. From the standpoint of environmental protection, a regulation of the problem as a whole would be preferable.

The advantage of agreements specifically targeted at sustainable tourism – at the global as well as regional levels – is that questions of sustainable tourism can be regulated directly and comprehensively. They are not individual aspects of other contexts but the main topic of the agreement.

The disadvantages of agreements dealing specifically with sustainable tourism are that the aspect of sustainability of tourism, so desirable from the environmental standpoint, could not be taken far enough, that other aspects could press to the fore and win the upper hand. Moreover, in new agreements in their own right all substantive and institutional questions must be negotiated from the beginning; it is not possible to build on existing agreements. That can be tedious and involve a loss of time. Besides, the political and institutional negotiation context for the new agreement must first be established. Finally, new and independent regulatory agencies mean additional costs for the states, which could be at odds with current trends in international diplomacy to prefer using existing regulatory agencies.

15.2 Additional Agreements (Protocols) to Existing Agreements

International regulations on sustainable tourism can also be adopted as additional agreements to existing international agreements. This is likewise possible at the global and regional levels. Agreements at varying levels have taken on increasing significance in recent years. Examples at the global level are the Convention for the Protection of the Ozone Layer (Vienna Convention of 1985; Montreal Protocol of 1987) and the still-to-be-expanded conventions on the protection of the climate and biological diversity. Examples at the regional level are the Geneva Convention on Long-Range Transboundary Air Pollution and the protocols attached to it, the Convention on the Protection of the Mediterranean or the Alps Convention and its protocols.

Agreements on sustainable tourism can be arrived at as additional agreements to existing agreements on nature conservation at the global and/or regional levels. Theoretically they are also possible as additional agreements to economically oriented agreements on tourism which have yet to be reached, but this is currently not the case and would also involve the risk that the sustainability of tourism could be an unheeded minor aspect. For this reason, the only currently relevant question is whether additional agreements to existing agreements on nature conservation should be worked out, and if so, which agreements these are.

The main advantage of additional agreements to existing conventions is that an existing negotiating context can be used and an existing substantive and institutional agreement can be built on. In addition to this practical advantage, the content of a new agreement is also already pointed in a certain direction through the convention it is based on.

Problems of additional agreements could at most be seen in the fact that attaching them to existing treaties already limits them in regard to the subject of the negotiations and the regulations to be created by them. But such objections are unfounded if comprehensive treaties are available which also leave enough room for negotiating the additional agreement.

For the problem of sustainable tourism this is the case. With the Convention on Biological Diversity a comprehensive agreement is available at the global level which can cover all aspects of sustainable tourism. Moreover, it leaves enough room for negotiating an additional agreement (cf. BURHENNE-GUILMIN 1993, p. 56; GÜNDLING 1995, p. 68). At the regional level, too, there are comprehensive agreements on nature conservation which could be a suitable framework for additional agreements on sustainable tourism. The Alps Convention is one such agreement for an important part of Europe. The Convention on the Mediterranean, which has already been considerably expanded, could also serve as a framework for an agreement on sustainable tourism. Also, all the other regional conventions discussed above have such potential; agreements on sustainable tourism in the regions they govern could be based on them.

15.3 Incorporating Additional Regulations into Existing Agreements

International agreements on sustainable tourism could also be reached by adding appropriate regulations to existing agreements on international nature conservation (or to other agreements, which is also a fundamental possibility). This would be the minimum solution, so to speak. Here, too, the advantage is the already existing substantive and institutional arrangement. But this would run the risk of not being able to regulate problems of sustainable tourism so immediately and comprehensively as would be the case in additional agreements.

15.4 Worldwide and/or Regional International Regulations on Sustainable Tourism?

It has already been stated several times that international regulations on sustainable tourism are possible at the global and regional levels, regardless of which option as to the form this regulation should have is chosen. The question now is at which level new international regulations are advisable, the global level, the regional one, or both.

This basically political question will not be answered with finality from the legal standpoint. But it may be pointed out that from the standpoint advocated here global as well as regional regulations are considered desirable. Global regulations are necessarily general because they must be capable of finding consensus among the global community of states. They can and should be given concrete form by agreements at the regional level which can go into the given conditions in the region in more detail, including the political conditions.

Although it is not absolutely imperative to begin with worldwide regulations, a time sequence of this kind could be advisable, after all. The worldwide agreement on sustainable tourism could constitute the framework for regional agreements, giving them a substantive direction and also political thrust for developing regional agreements.

15.5 Political Directives to Pave the Way for Legal Regulations on Sustainable Tourism

The practise of international legislation shows that often the way for legal regulations is prepared politically and substantively by legally non-binding instruments. In particular, in areas which are regarded as politically sensitive, this is often observed to be the case. Tourism can definitely be classified as being such an area: thus far it has hardly been regulated, which should be attributed to conscious reluctance and not to negligence towards the problem. Accordingly, caution could be called for when introducing legal regulations. Directives containing the major elements of possible international agreements, but which do not yet make them mandatory, could serve to pave the way for this effort.

15.6 Assessment of Options

From the standpoint of legal strategy, all the options named above are open in principle. International regulations on sustainable tourism can be worked out as agreements in their own right as well as additional agreements to existing conventions on nature conservation and in particular, at the global level, as a protocol to the Convention on Biological Diversity.

In an independent agreement which is not attached to the Convention or another agreement, tourism can also be regulated according to the standpoint of sustainability. An agreement in its own right need not necessarily be "weaker" in regard to environmental and nature-protection criteria, for an agreement such as this could certainly be given the provisions of a specific environmental and nature-conservation agreement. But the danger cannot be overlooked that sustainability could end up not being its main aspect.

The questions of assessing the options are essentially political. However, a number of reasons can be suggested for attaching legal instruments on sustainable tourism to agreements on nature conservation, for choosing the global as well as the regional levels for such endeavours and for beginning with the global level and as a first step drawing up a protocol to be attached to the Convention on Biological Diversity.

The main idea of sustainable tourism is the protection and preservation of nature as the basis of tourism. The opposite side of the coin is that tourism can impact nature in a massive way if it is not regulated. Tourism is thus a major factor which nature-conservation efforts must address.

If a legal instrument is attached to an existing nature-conservation agreement, its fundamental ideas can be further built on. The principles are already there, so they need not be negotiated again starting from scratch.

The same is true in regard to procedure. When attached to an existing agreement, the framework is already there within which the legal instrument on sustainable tourism is to be negotiated. An already existent agreement provides the institution and procedure in which the text can be prepared up to its final version, which can then be adopted. Together with the substantive provisions, this affords the assurance that a legal instrument on sustainable tourism can be drawn up.

Attaching legal regulations concerning sustainable tourism to nature-conservation agreements is possible and desirable at the global as well as regional levels. Regional agreements can complement global ones and give them more concrete form. It would be most appropriate to begin with global agreements.

An argument for attaching worldwide regulations on sustainable tourism to the Convention on Biological Diversity (as an additional protocol to this

Convention) is that the Convention has become the crystallisation point of nature conservation worldwide. The Convention has placed global nature protection on new foundations. It pursues a systematic approach and has made the main idea of all nature conservation the object of protection: the preservation of biological diversity, taken comprehensively as the diversity of ecosystems, diversity of species and genetic diversity. The Convention builds on existing agreements and in principle provides that they remain in effect (Article 22), but points them in the direction of the idea of protecting biological diversity. It is designed to have further specific additional protocols devoted to special problems of the protection and preservation of biological diversity. Additional protocols can also deal with the regulation of certain processes and activities which impact the preservation of biological diversity, such as touristic activities.

Therefore the position is advocated here that if tourism is considered to be in need of regulation in regard to sustainability and if further instruments of international law are deemed necessary to achieve such regulation, an additional protocol to the Convention on Biological Diversity would be the appropriate way to begin with the formulation of the required legal regulations.

Before drafting and proposing such an additional protocol, the question of the level of detail of a worldwide agreement on sustainable tourism will be discussed.

16 On the Level of Detail of a Worldwide Agreement on Sustainable Tourism

The level of detail of an agreement is a statement as to which subjects exactly are treated in an agreement and to what extent of concreteness.

For a global instrument on sustainable tourism drawn up as a protocol additional to the Convention on Biological Diversity, the following considerations, among others, will probably be relevant:

Like the Convention itself it would be a global protocol. It must therefore be drawn up in such a way that it is acceptable in principle by all states. That means that it must limit itself to questions of general significance and must leave room for states and the affected international organisations to have their various interests and positions included to a certain extent, while at the same time preserving a minimum of control and guidance. An international agreement to which the members of the international community voluntarily accede, in keeping with the principle of sovereignty, must be realistically drawn up, and that means that there must be a balance between the demands and the leeway which make participation acceptable to as many states and international organisations as possible. Regarding the level of detail, it should also be borne in mind that thus far the states have been reluctant to create legally binding regulations on sustainable tourism. This consideration would serve to prevent overreaching the demands on the contracting parties, but without creating all-too "soft" provisions or, worse, mandatory ones devoid of any substance.

An additional protocol to the Convention on Biological Diversity can build on a number of substantive and possibly also institutional provisions of the Convention. They would not have to be negotiated and set forth in a protocol to this effect. It would always need to be weighed as to what extent there are particular problems apart from the fundamental regulations of the Convention which result from the special problem of sustainable tourism.

17 Proposal for a Worldwide Regulation Concerning Sustainable Tourism as a Protocol Additional to the Convention on Biological Diversity

Protocol on Sustainable Tourism

Preamble

The Parties of this Protocol,

- acknowledging the growing economic significance of tourism;
- in view of the worldwide dimensions of modern tourism, particularly the steadily increasing numbers in the sector of long-distance tourism;
- aware that international steps and regulations are called for;
- cognisant of the fact that an intact nature and the diversity of life are the foundations of many touristic activities;
- in recognition of the fact that sustainable tourism can contribute to the protection and sustainable use of biological diversity;
- concerned that uncontrolled tourism endangers nature and biological diversity as foundations of tourism and in some regions has already damaged it;
- mindful that touristic activities require social and political control, particularly through self-regulatory steps of the tourism industry and through planning instruments, economic incentives, awareness-building programmes and training measures by the political sector to ensure sustainability;
- resolved to devote particular attention to touristic activities in especially fragile areas, namely in protected areas, mountain regions, coastal areas and other attractive nature areas;
- considering that in the promotion of sustainable tourism the interests of the local communities including indigenous peoples must be borne particularly in mind and
- in recognition that traditional cultures deserve to be protected,

agree as follows:

Part I: General Provisions

Article
Objective

The objective of this Protocol is the support of sustainable tourism, which, guided by the principle of environmentally sustainable development, bears the requirements of nature and environmental protection and in particular the preservation of biological diversity in mind.

Article
Definitions

Inasmuch as in this Protocol terms are used which are defined in the Convention on Biological Diversity, the definitions of the Convention apply.

For the purposes of this Protocol

"indigenous population" means that group of the population in the territory of a Party which was residing there when the state was founded and which preserves some or all of its traditional social, economic, cultural and political institutions;

"Convention" means the Convention on Biological Diversity;

"ecological balances" means the systematic identification and assessment of resource consumption and the environmental stress caused by carrying out an activity or manufacturing a product;

"ecological carrying capacity" means the capacity of nature and its resources to compensate for exploitation and stress and to preserve the potential to fulfill the requirements and desires of the present and future generations;

"economic incentives" means financial incentives, in particular subsidies, tax exemptions, tax relief and rewards to promote sustainable behaviour and financial burdens, in particular fees and taxes to reduce or avoid environmentally adverse or damaging behaviour;

"Secretariat" means the Secretariat of the Convention;

"tourism" means the sum total of activities in connection with recreation and leisure time, including recreational and leisure-time activities, the services offered for recreation and leisure time and the providing of infrastructure for recreation and leisure time;

"sustainable tourism" means tourism which makes the requirements of nature and environmental protection the basis for touristic activities, thus ensuring its ecological sustainability;

"environmental-impact assessment" means the assessment provided for in Article 14 of the Convention.

Part II: Basic Obligations

Article
Obligation to Promote Sustainable Tourism

The Parties shall promote sustainable tourism and protect, preserve and develop nature and the environment as a foundation of tourism, as well.

The Parties shall take appropriate steps to develop tourism in a way which is mindful of the requirements of nature and environmental protection and the preservation of biological diversity. Wherever nature is used for touristic purposes the steps shall ensure that the ecological carrying capacity is not exceeded and that nature is sustained as the foundation of tourism.

Article
Precautionary Steps and Monitoring

The Parties shall take preventive steps to preclude damage to nature and biological diversity. Existing touristic activities shall be monitored as to their environmental impact in order to ensure sustainability.

Article
Support of Nature Tourism

The Parties shall support such forms of nature tourism which are geared in a sustainable way to enjoying special nature experiences and as such make a contribution to the preservation and sustainable use of biological diversity.

Article
International Co-operation

Inasmuch as it is required and appropriate, the Parties shall co-operate internationally – individually or within the framework of international organisations – to stipulate the programmes and regulations necessary for the execution of the general obligations.

The Parties shall make an effort to promote tourism in their countries and internationally in such a way that it can make a contribution to the fair development of the global economy.

Part III: Steps for Supporting Sustainable Tourism

Article
Surveys

The Parties shall submit surveys of touristic activities in their sovereign territory and include in these surveys the ecological impacts of existing touristic activities.

The Parties shall determine criteria for the stress on nature and the environment stemming from tourism. They shall designate those areas in which the limits of the carrying capacity have been reached or exceeded (areas under stress).

The Parties shall use ecological balances for touristic activities. They shall co-operate internationally to develop common criteria and procedures for statistical surveys.

Article
Promotion of Quality

The Parties shall ensure a sustainable quality of tourism and be mindful of the security and health of the tourists to an appropriate degree.

Article
Environmentally Oriented Tourism Planning

The Parties shall plan touristic activities and include in the planning processes ecological requirements and possible impacts (environmentally oriented tourism planning). Tourism plans shall be integrated in development plans in such a way that development planning bears the environmentally oriented tourism planning in mind. When planning touristic developments in areas close to national borders, the Parties on both sides of the borders shall co-operate.

The Parties shall see to it that environmentally oriented tourism planning is implemented and that the implementation is monitored.

Article
Assessment of Environmental Impacts

Touristic activities, including tourism planning, steps providing touristic infrastructure and other activities relevant to tourism shall be subject to an environmental-impact assessment if significant impacts on the environment may result from them.

In the case of activities which have relevance beyond the national borders, the Parties shall co-operate internationally in the environmental-impact assessment, inter alia through mutual information and consultation of government agencies and the public.

Article
Tourism and Transportation

Touristic developments shall be designed on the basis of sustainable transportation programmes. If need be, steps to reduce the traffic load shall be taken, in particular to reduce motorised traffic.

The Parties shall take all necessary steps to keep the stress from air traffic at an acceptable level. Inasmuch as existing air traffic is no longer within the limits of acceptability, the Parties shall take steps to reduce the stress stemming from air traffic.

Article
Sports Activities

Sports activities by tourists shall be regulated in such a way that they are in keeping with the requirements of nature, landscape and environmental protection.

Tourist sports activities which are not in keeping with the requirements of sustainable tourism shall be prohibited.

Article
Hunting by Tourists

The Parties shall control the hunting activities of tourists and ensure that the concerns of nature conservation, in particular species protection, are respected.

Article
Support of the Local Economy

The Parties shall make the interests of the local populace and the structures of the local economy the basis of their programmes and steps in support of sustainable tourism.

In this connection the Parties shall take all appropriate steps to bolster the local economy through sustainable tourism.

Article
Indigenous Population

The Parties shall bear the interests of the indigenous population in mind in promoting sustainable tourism. They shall ensure the protection and preservation of these interests on an economic, social and cultural basis.

Indigenous groups of the population shall be involved in the planning of touristic activities in an effective way wherever they are directly affected by the touristic activities. Care shall be taken that they are given the share they deserve of the benefits derived from the touristic development.

Inasmuch as protective provisions to ensure the sustainability of touristic activities lead to restrictions for indigenous population groups, the Parties shall promote the development of occupational alternatives. In severe cases economic losses shall be compensated for.

Article
Protection of Traditional Cultures and Life-styles

The Parties shall take the necessary steps to preserve traditional cultures and life-styles.

Article
Economic Incentives

The Parties shall provide for the use of economic incentives, where appropriate and effective. They shall encourage and promote voluntary control mechanisms established by the tourism industry to ensure sustainable tourism.

Article
Co-operation at All Levels

In all steps which serve to ensure sustainable tourism, the Parties shall promote co-operation at all levels of government as well as the co-operation of governmental and non-governmental institutions.

Part IV: Tourism in Special Areas

Article
Tourism in Protected Areas

The Parties shall control and regulate tourism in protected areas in co-operation with the institutions responsible for the protected areas in ways which are mindful of the requirements of the protected areas.

The Parties shall ensure that tourism in protected areas is environmentally sustainable and does not counteract the objectives of conservation.

The Parties shall keep particularly fragile protected areas, reserves located in protected areas and all other protected areas free from tourism of all kinds if the objectives of protection so require.

The management of protected areas and the control of visitors through qualified personnel shall be ensured.

Article
Tourism in Areas Under Stress

Areas which are designated as areas under stress in accordance with Article ... shall be kept free from further touristic activities. The Parties shall reduce existing stress in these areas and introduce steps to restore them.

In these cases the Parties shall take all steps to promote the development of alternative economic activities.

Article
Tourism in Coastal Areas

The Parties shall take the necessary preventive and monitoring steps in coastal areas to ensure sustainable tourism. The internationally developed principles of integrated coastal-zone management shall be borne in mind.

The Parties shall take all necessary steps to protect sensitive areas, including islands, coral reefs, coastal ocean waters, mangroves, coastal wetlands, sandy beaches and coastal dunes.

Article
Tourism in Mountain Areas

The Parties shall ensure sustainable tourism in mountain areas.

Mass tourism in sensitive zones of mountain areas shall be avoided.

The Parties shall in particular control and regulate tourism in high mountains in such a way that it is in keeping with the conditions for preserving the biological diversity of these areas.

Inasmuch as sustainable tourism in mountain areas contributes to an economic upswing in these areas and can counteract the emigration of the local populace, the Parties shall do everything they can to support such forms of tourism.

Article
Tourism in Nature Areas

The Parties shall manage tourism in all areas which are characterised by rich and attractive natural scenery in a sustainable way. In doing so they shall pay special heed to the protective and preservation requirements of forest areas, freshwater ecosystems, spectacular nature areas and arctic and antarctic ecosystems.

Article
More Stringent National Provisions

To ensure sustainable tourism the Parties may take more stringent steps at the national level than provided for in this Protocol.

Part V: Training, Awareness-Building, Research

Article
Training and Awareness-Building

The Parties shall take steps to incorporate environmental aspects in the training of the people employed in tourism. Similarly, they shall inform the public on the requirements of sustainable tourism.

Article
Research

The Parties shall promote research on sustainable tourism and facilitate the international exchange of knowledge in this field.

Part VI: Control and Implementation

Article
Control and Implementation at the National Level

The Parties shall establish control and monitoring steps in order to ensure that the steps taken are in keeping with this Protocol. If necessary and appropriate, they shall co-operate at the international level and within the framework of international organisations.

Article
Reports by the Parties

The Parties shall report regularly to the Secretariat on the steps taken in keeping with this Protocol. The reports shall also discuss the effectiveness of these steps.

The Conference of the Parties shall determine the time for the presentation of the first report and the intervals for the presentation of further reports.

The Secretariat shall review the reports of the Parties and in turn submit its report to the Conference of the Parties.

The Conference of the Parties shall take note of the reports of the parties. In cases where obligations have not been kept, the Conference may make recommendations to alleviate the state of non-compliance.

Article
Inspection of the Effectiveness of Protocol Provisions

The Parties shall regularly inspect the provisions set forth in this Protocol as to their effectiveness. If need be they shall bring about the necessary amendments to this Protocol.

Part VII: Institutional Provisions

Article
Relationship of This Protocol to the Convention

In accordance with Article 32 of the Convention, only Contracting Parties of the Convention may become Parties to this Protocol.

Article
Resolutions

Resolutions on the basis of this Protocol shall be adopted by the Parties of this Protocol. Contracting Parties of the Convention which did not ratify, adopt or approve this Protocol may attend each meeting of the Parties as observers.

Article
Tasks of the Secretariat

The secretarial tasks required for carrying out this Protocol shall be performed by the Secretariat.

Article
Expert Advice

The Subsidliary Body on Scientific, Technical and Technological Advice of the Convention shall assume the task of expert advice for purposes of implementing this Protocol. Advice shall be given upon request of the Parties and in accordance with the Convention and the criteria stipulated under the Convention.

Part VIII: Final Provisions

Article
Relationship of This Protocol to Other International Agreements

This Protocol shall not affect the rights and obligations of the Parties under existing international agreements insofar as they do not contradict the goals and provisions of this Protocol.

Article
Signature

This Protocol shall be open for signature by all states and organisations of regional economic integration in ... from ... to ...

Article
Ratification, Adoption or Approval

This Protocol shall require the ratification, adoption or approval by the states and organisations of regional economic integration. The instruments of ratification, adoption or approval shall be deposited with the depositary.

Article
Accession

From that day forward when this Protocol is no longer open for signature, states and organisations of regional economic integration may accede to it. The instruments of accession shall be deposited with the depositary.

Article
Entry into Force

This Protocol shall enter into force ninety days after the ...th instrument of ratification, adoption, approval or of accession has been deposited.
The time of entry into force for each Party shall be determined by Article 36, section 4 of the Convention.

Article
Reservations

No reservations may be made to this Protocol.

Article
Withdrawals

After a period of two years after the time when this Protocol has entered into force for it, a Party may withdraw from this Protocol by addressing notification to the depositary of this Protocol. The withdrawal shall enter into force one year after receipt of the notification by the depositary or at a later time specified in the notification of withdrawal.

The withdrawal from this Protocol does not affect the rights and obligations from the Convention of the Party withdrawing.

Article
Depositary

The Secretary-General of the United Nations shall assume the task as depositary of this Protocol.

Article
Authentic Texts of This Protocol

The Arabic, Chinese, English, French, Russian and Spanish texts of this Protocol are equally authentic.

Done at ..., on ...

18 Conclusion

"We need not invent new steps, but just use existing instruments and imaginatively combine them. There is enough theoretical knowledge and there are instruments of control, too – the problem really is their political implementation."

KRIPPENDORF 1982

This statement by Krippendorf must be scrutinised a bit more carefully in the light of the preceding investigations. There are indeed implementation problems for the (few) existing control instruments; but there is also clearly a need for introducing new instruments and gearing them specifically to ensure a tourism which is sustainable for nature and the environment.

The goal of a tourism which is sustainable for nature and the environment does not primarily consist of establishing restrictions for touristic activities. The goal is rather – and this should be stressed in closing – to preserve nature as the foundation of tourism and to enable this form of nature use to be lasting.

The portions of this study dealing with nature conservation as such show that solution strategies have been developed as plans, but are thus far without any concrete relevance to tourism. Only in the programmes and agreements of the states bordering the Mediterranean is the environmental stress on the coasts a part of the concrete plan of action MAP (Mediterranean Action Plan). To what extent the various instruments in the countries of the Mediterranean region have really been effective, has yet to be studied. What is needed are detailed effectiveness analyses, but these are not yet available.

The result in the legal sector is similar. There are no special regulations to ensure a tourism sustainable for nature and the environment in laws existing today. Such regulations are currently being negotiated, and at that only at the regional level (Alpine Region). Megaregional or indeed global regulations have thus far not been attempted.

This situation will not do justice in the long run to the problems which exist due to and for tourism. The need for control is unavoidable, for the sake of tourism, too. Whether legally binding agreements should be created right from the outset – in whatever form, as additional agreements (protocols) to existing treaties or in other form – is another question.

It would be conceivable to lead the way with political directives to prepare the ground for a consensus on legal regulations. As with all solutions involving directives, this also entails the risk that urgently required steps could be put off indefinitely. This danger can be combatted by drawing up the directives speedily and with particular commitment. It is also important that the directives be followed, despite their political character.

The adoption of the Charter for Sustainable Tourism at the conclusion of the World Conference on Sustainable Tourism in Lanzarote in 1995 can be viewed as a first step. The principles adopted there should get a discussion going to bolster the same or similar principles and thus point in the direction of forming a consensus.

The study also illustrates that hitherto by far not all of the existing solution strategies have been tried *(cf. Table 24)*. There is a deficit of implementation and execution which exists in regard to political steps and legal regulations and which must be expressly addressed. Existing solutions must be implemented; with new ones, more effective implementation must be part of the planning.

It is not enough to come up with new control instruments and then leave them to their own devices, i.e., to being ignored. Implementation and execution must be strengthened, and for this a whole range of steps is required, among which is achieving a broad awareness of the problems of tourism, but also of the existing instruments to incorporate tourism in formal education and involving nature-conservation and environmental-protection groups in administrative and court proceedings.

This shows that new programmes and also new international agreements must not simply be new versions of already extant contents. What is needed instead is that problems of sustainable tourism be attacked specifically and concretely. New sets of instruments, regardless of what kind, must not be limited to generally worded objectives which leave the responsible institutions plenty of room to decide if they want to act and how they should. Agreeing on concrete steps to ensure the sustainability of tourism is the only thing which makes sense.

The fundamental precondition for developing and implementing effective control mechanisms is that at the international as well as national levels there is a realisation that tourism constitutes a threat to the preservation of biological and landscape diversity and that specific solution strategies are needed to avoid and reduce conflicts.

Appendix E. Tourism statistics for individual countries[1]

Tourism statistics, North America

	Arrivals 1992	Arrivals 1988	Change	Land area	Arrivals 1992	Foreign-currency income from tourism
	(1,000)	(1,000)	(%)	(sq. km.)	(per sq. km. of land area)	(%) 1992
U.S.A.	44,647	34,095	+31	9,363,000	4.8	10.7
Mexico	17,271	14,142	+22	1,973,000	8.8	17.9
Canada	14,741	15,485	-5	9,976,000	1.5	4.1

[1] Source: calculations based on data in WTO 1994a.

Tourism statistics, Central America

	Arrivals 1992	Arrivals 1988	Change	Land area	Arrivals 1992	Foreign-currency income from tourism
	(1,000)	(1,000)	(%)	(sq. km.)	(per sq. km. of land area)	(%) 1992
Costa Rica	610	329	+85	51,000	12.0	24.8
Guatemala	541	405	+34	108,000	5.0	18.7
El Salvador	314	134	+134	21,000	15.0	7.6
Panama	291	199	+46	77,082	3.8	28.9
Belize	247	142	+74	22,963	10.8	43.4
Honduras	230	162	+42	112,088	2.1	2.9
Nicaragua	167	55	+204	118,558	1.4	8.5

Tourism statistics, South America

	Arrivals 1992	Arrivals 1988	Change	Land area	Arrivals 1992	Foreign-currency income from tourism
	(1,000)	(1,000)	(%)	(sq. km.)	(per sq. km. of land area)	(%) 1992
Argentina	3,031	2,119	+43	2,766,889	1.1	20.2
Uruguay	1,802[1]	1,036[1]	+74[1]	176,000	10.2[1]	18.3
Brazil	1,475	1,743	-15	8,511,965	0.2	3.5
Chile	1,283	624	+106	757,626	1.7	6.6
Columbia	1,076	829	+30	1,141,568	0.9	6.0[2]
Venezuela	434	373	+16	912,000	0.5	3.2
Ecuador	403	347	+16	272,045	1.5	6.0
Paraguay	334	284	+18	406,752	0.8	18.9
Bolivia	245	167	+47	1,098,581	0.2	12.5
Peru	217	359	-40	1,285,215	0.2	6.9[2]
Guyana	93[1]	71[1]	+17[1]	214,969	0.4[1]	9.8
Suriname	30	21	+43	163,820	0.2	2.3[3]

[1] including daytime tourists

[2] 1991

[3] 1990

Tourism statistics, Caribbean

	Arrivals 1992	Arrivals 1988	Change	Land area	Arrivals 1992	Foreign-currency income from tourism
	(1,000)	(1,000)	(%)	(sq.km.)	(per sq.km. of land area)	(%) 1992
Puerto Rico	2,640	2,281	+16	8,900	296.6	6.2[1]
Dominican Republic	1,524	1,116	+37	49,520	30.8	65.9
Bahamas	1,399	1,475	-5	13,942	100.3	44.0[1]
Jamaica	909	649	+40	11,425	79.6	42.9
Saint Maarten (Netherl. Antilles)	569	480	+19	41	13,878.0	?
Aruba	542	278	+95	193	2,808.3	90.0[2]
U.S. Virgin Islands	487	556	-12	342	1,424.0	25.6
Cuba	408[1]	298	+37	110,860	3.7[1]	7.4[1]
Barbados	385	451	-15	431	893.3	70.2
Bermuda	375	427	-12	53	7,075.5	84.1
Martinique	321	280	+15	1,080	297.2	53.3
Guadeloupe	300	329	-9	1,800	166.7	67.4
Cayman Islands	242	219	+11	260	930.8	98.7[1]
Trinidad and Tobago	235	186	+26	5,128	45.8	5.6
Antigua and Barbuda	210	187	+12	440	477.3	91.8[2]
Curaçao (Netherl. Antilles)	207	155	+34	544	380.5	?
Saint Lucia	177	133	+33	616	287.3	62.8
Haiti	120	133	-10	27,750	4.3	30.9[1]
British Virgin Islands	117	176	-34	150	780.0	?
Saint Kitts and Nevis	90	70	+29	269	334.6	?
Grenada	88	59	+49	344	255.8	63.6[1]

[1] 1991

[2] 1988

Tourism statistics, Caribbean (continued)

	Arrivals 1992 (1,000)	Arrivals 1988 (1,000)	Change (%)	Land area (sq.km.)	Arrivals 1992 (per sq.km. of land area)	Foreign-currency income from tourism (%) 1992
St. Vincent and Grenadines	53	47	+13	388	136.6	7.8[3]
Turks and Caicos Is.	52	47	+11	417	124.7	?
Bonaire (Netherl. Antilles)	52[4]	34[4]	+53[4]	311	167.2[4]	?
Dominica	47	34	+38	750	62.7	26.7[3]
Anguilla	32	28	+14	91	351.6	?
St. Eustatius (Netherl. Antilles)	30[4]	18[4]	+67[4]	21	1,428.6[4]	?
Saba (Netherl. Antilles)	25[4]	25[4]	0	13	1,923.1[4]	?
Montserrat	17	18	-6	102	166.7	91.7[5]

[3] 1990

[4] including daytime tourists

[5] 1989

Tourism statistics, North Africa

	Arrivals 1992	Arrivals 1988	Change	Land area	Arrivals 1992	Foreign-currency income from tourism
	(1,000)	(1,000)	(%)	(sq.km.)	(per sq.km. of land area)	(%) 1992
Morocco	4,390	2,841	+55	710,850	6.2	25.5
Tunesia	3,540	3,468	+2	164,150	21.6	21.0
Algeria	1,120	967	+16	2,381,741	0.5	0.9[1]
Sudan	16[1]	37	-57	2,505,813	0[1]	6.4[2]

[1] 1991

[2] 1990

Tourism statistics, East Afrika

	Arrivals 1992	Arrivals 1988	Change	Land area	Arrivals 1992	Foreign-currency income from tourism
	(1,000)	(1,000)	(%)	(sq.km.)	(per sq.km. of land area)	(%) 1992
Zimbabwe	738	449	+64	390,759	1.9	7.2[1]
Kenya	699	695	+1	582,646	2.0	5.0
Mauritius	335	239	+40	1,865	179.6	18.8
Réunion	217	182[2]	+19	2,512	86.4	?
Tanzania	202[3]	130[3]	+55[3]	945,050	0.2[3]	20.9[4]
Zambia	159	108	+47	752,614	0.2	0.4[4]
Malawi	123[3,4]	99[3]	+24[3]	118,480	1.0[3]	2.7[4]
Seychelles	99	77	+29	455	217.6	19.7[4]
Burundi	86	99	-13	27,834	3.1	5.1
Ethiopia	83	76	+9	1,221,900	0.1	9.6[4]
Uganda	69[4]	40	+73	236,600	0.3[4]	6.9[4]
Madagascar	54	35	+54	587,041	0.1	8.1[4]
Djibouti	33[4]	29	+14	23,200	1.4[4]	?
Comoro Isl.	19	8	+138	2,235	8.5	?
Ruanda	5	19[2]	-74	26,338	0.2	5.6

[1] 1990

[2] 1989

[3] including daytime tourists

[4] 1991

Tourism statistics, Southern Africa

	Arrivals 1992	Arrivals 1988	Change	Land area	Arrivals 1992	Foreign-currency income from tourism
	(1,000)	(1,000)	(%)	(sq.km.)	(per sq.km. of land area)	(%) 1992
South Africa	2,892	805	+386	1,221,037	2.4	5.3
Botswana	412[1]	268	+54	581,730	0.7[1]	?
Swaziland	258	196	+32	17,364	14.9	
Lesotho	155	110	+41	30,355	5.1	?
Namibia	?	166	?	824,292	0.2[2]	?

[1] 1991
[2] 1988

Tourism statistics, Central Africa

	Arrivals 1992	Arrivals 1988	Change	Land area	Arrivals 1992	Foreign-currency income from tourism
	(1,000)	(1,000)	(%)	(sq.km.)	(per sq.km. of land area)	(%) 1992
Gabun	128	20	+540	267,667	0.5	0.1[1]
Cameroons	62	100	-38	475,442	0.1	1.1[2]
Angola	55[2,3]	?	?	1,246,700	0[2,3]	0.3[1]
Congo	37	39	-5	342,000	0.1	0.8[2]
Zaire	22	39	-44	2,344,858	0	0.7[1]
Chad	21	21	0	1,284,000	0	4.7[2]

[1] 1990
[2] 1991
[3] including daytime tourists

Tourism statistics, West Africa

	Arrivals 1992	Arrivals 1988	Change	Land area	Arrivals 1992	Foreign-currency income from tourism
	(1,000)	(1,000)	(%)	(sq.km.)	(per sq.km. of land area)	(%) 1992
Senegal	246	256	-4	196,722	1.3	?
Nigeria	237	143	+ 66	923,768	0.3	0.2
Ivory Coast	217	178	+22	323,000	0.7	2.2[1]
Ghana	213	114	+87	238,537	0.9	6.6[1]
Benin	130	75	+73	112,622	1.2	?
Gambia	114[2]	102	+12	11,295	10.1[2]	56.1[2]
Sierra Leone	91	75	+21	72,326	1.3	9.3
Burkina Faso	74	83	-11	274,000	0.3	5.9[1]
Togo	49	104	-53	56,785	0.9	16.2[2]
Mali	38	36	+6	1,240,142	0	12.6[1]
Niger	13	33	-61	1,267,000	0	5.7[3]

[1] 1989
[2] 1991
[3] 1990

Tourism statistics, Northeast Asia

	Arrivals 1992	Arrivals 1988	Change	Land area	Arrivals 1992	Foreign-currency income from tourism
	(1,000)	(1,000)	(%)	(sq.km.)	(per sq.km. of land area)	(%) 1992
China	16,500[1]	1,842[2]	+117[2]	9,600,000	1.7	4.5
Hong Kong	6,986[3]	5,589[3]	+25[3]	1,068	6,541.2[3]	4.8
South Korea	3,231[3]	2,340[3]	+38[3]	99,143	32.6[3]	4.1
Macao	3,180	2,513[4]	+27	17	187,058.8	56.1
Japan	2,103	1,116	+88	377,835	5.6	1.0
Taiwan	1,873	1,935	-3	36,179	51.8	?

[1] revised figure, WTO News 1/1994
[2] original figures
[3] including daytime tourists
[4] 1990

Tourism statistics, Southeast Asia

	Arrivals 1992	Arrivals 1988	Change	Land area	Arrivals 1992	Foreign-currency income from tourism
	(1,000)	(1,000)	(%)	(sq.km.)	(per sq.km. of land area)	(%) 1992
Malaysia	6,016	3,624	+66	330,000	18.2	4.2
Singapore	5,446	3,833	+42	633	8,603.5	7.6
Thailand	5,136	4,231	+21	514,000	10.0	13.1
Indonesia	3,064	1,301	+136	1,919,400	1.6	7.9[1]
Philippines	1,043	1,023	+2	300,000	3.5	14.6
Brunei Darussalam	500	457	+9	5,765	86.7	1.7[2]

[1] 1991
[2] 1989

Tourism statistics, Australia and New Zealand

	Arrivals 1992	Arrivals 1988	Change	Land area	Arrivals 1992	Foreign-currency income from tourism
	(1,000)	(1,000)	(%)	(sq.km.)	(per sq.km. of land area)	(%) 1992
Australia	2,603[1]	2,249[1]	+16[1]	7,687,000	0.3[1]	8.6
New Zealand	1,056	865	+22	269,000	3.9	9.5

[1] including daytime tourists

Tourism statistics, Antarctica

	Arrivals 1992	Arrivals 1988	Change	Land area	Arrivals 1992	Foreign-currency income from tourism
	(1,000)	(1,000)	(%)	(sq.km.)	(per sq.km. of land area)	(%) 1992
Antarctica	6,6	5,0	+32	5,500,000	0.0	–

Sources: SPLETTSTOESSER 1994, SMITH 1994

Tourism statistics, Melanesia

	Arrivals 1992	Arrivals 1988	Change	Land area	Arrivals 1992	Foreign-currency income from tourism
	(1,000)	(1,000)	(%)	(sq.km.)	(per sq.km. of land area)	(%) 1992
Fiji Islands	279	208	+34	18,333	15.2	33.9
New Caledonia	78	61	+28	18,575	4.2	18.7
Vanuatu	43	16	+169	14,760	2.9	55.6[1]
Papua-New Guinea	41	41	0	462,840	0.1	2.6
Solomon Islands	12	11	+9	27,556	0.4	5.4[2]

[1] 1991

[2] 1990

Tourism statistics, Micronesia and Hawaii

	Arrivals 1992	Arrivals 1988	Change	Land area	Arrivals 1992	Foreign-currency income from tourism
	(1,000)	(1,000)	(%)	(sq.km.)	(per sq.km. of land area)	(%) 1992
Hawaii	6,514	6,142	+6	16,760	388.7	?
Guam	877	586	+50	541	1,621.1	?
Northern Mariana Is.	505[1]	246[1]	+105[1]	477	1,058.7[1]	?
Marshall Islands	8	4	+100	?	?	?
Kiribati	4[1]	3[1]	+33[1]	717	5.6[1]	?

[1] including daytime tourists

Tourism statistics, Polynesia

	Arrivals 1992	Arrivals 1988	Change	Land area	Arrivals 1992	Foreign-currency income from tourism
	(1,000)	(1,000)	(%)	(sq.km.)	(per sq.km. of land area)	(%) 1992
French Polynesia	124	135	-8	4,000	31.0	64.2
Cook Is.	50	34	+47	240	208.3	?
Samoa	38	49	-22	2,830	13.4	76.0
American Samoa	31	39	-21	29,000	1.1	3.0[1]
Tonga	23	19	+21	699	32.9	40.9
Niue	2	1	+100	185	10.8	?
Tuvalu	1	1	0	24	41.7	?

[1] 1991

Tourism statistics, South Asia

	Arrivals 1992	Arrivals 1988	Change	Land area	Arrivals 1992	Foreign-currency income from tourism
	(1,000)	(1,000)	(%)	(sq.km.)	(per sq.km. of land area)	(%) 1992
India	1,868	1,591	+17	3,288,000	0.6	7.3
Sri Lanka	394	183	+115	65,610	6.0	7.4
Pakistan	352	460	-23	804,000	0.4	1.6
Nepal	334	266	+26	147,181	2.3	22.7
Maldives	236	156	+51	298	791.9	73.9
Iran	212[1]	67	+216	1,648,000	0.1[1]	?
Bangladesh	110	121	-9	144,000	0.8	0.5[1]
Myanmar	21[1]	26	-19	677,129	0[1]	3.1[1]
Bhutan	3	2	+50	44,600	0.1	?

[1] 1991

Tourism statistics, Middle East

	Arrivals 1992	Arrivals 1988	Change	Land area	Arrivals 1992	Foreign-currency income from tourism
	(1,000)	(1,000)	(%)	(sq.km.)	(per sq.km. of land area)	(%) 1992
Egypt	2,944	1,833	+61	1,001,449	2.9	46.8
Bahrain	1,419	1,171	+21	598	2,372.9	4.5[1]
Dubai	944	598	+58	3,900	242.1	?
Saudi Arabia	750	763	-2	2,200,000	0.3	6.7[2]
Syria	684	421	+62	185,180	3.7	16.2
Jordan	661	608	+9	98,000	6.7	27.9
Iraq	504[3]	1,209[3]	-58[3]	434,924	1.2[3]	?
Oman	192	126	+52	300,000	0.6	1.5
Qatar	141	113	+25	11,437	12.3	?
Libya	89	98	-9	1,759,540	0.1	0[4]
Yemen	72	60	+20	195,000	0.4	?
Kuwait	65	80	-19	18,000	3.6	3.9

[1] 1991

[2] 1989

[3] including daytime tourists

[4] 1990

Tourism statistics, Central and Eastern Europe

	Arrivals 1992	Arrivals 1988	Change	Land area	Arrivals 1992	Foreign-currency income from tourism
	(1,000)	(1,000)	(%)	(sq.km.)	(per sq.km. of land area)	(%) 1992
Hungary	20,188	10,563	+91	93,030	217.0	10.5
Poland	13,500	11,350[1]	+19	312,683	43.2	23.7
Rumania	6,280[2]	5,514[2]	+14[2]	237,500	26.4[2]	6.1
Bulgaria	6,124[2]	8,295[2]	-26[2]	111,000	55.2[2]	1.2
former USSR	2,235[1]	2,458	-9	22,402,000	0.1	?
Slovakia	579	855	-32	49,030	11.8	5.7
Estonia	372	658[3]	-43	45,100	8.2	?

[1] 1991

[2] including daytime tourists

[3] 1989

Tourism statistics, Northern Europe

	Arrivals 1992	Arrivals 1988	Change	Land area	Arrivals 1992	Foreign-currency income from tourism
	(1,000)	(1,000)	(%)	(sq.km.)	(per sq.km. of land area)	(%) 1992
Great Britain	17,225[1]	14,939[1]	+16[1]	244,046	70.7[1]	6.7
Denmark	11,100[1]	7,546[1]	+47[1]	43,069	257.7[1]	8.7
Sweden	5,438[1]	6,662[1]	-18[1]	449,964	12.1[1]	5.2
Ireland	3,666	3,007	+22	70,280	52.2	5.4
Norway	2,375	1,704	+39	324,000	7.3	5.3
Iceland	143	129	+11	103,000	1.4	7.8

[1] unclear if daytime tourists included

Tourism statistics, Western Europe

	Arrivals 1992	Arrivals 1988	Change	Land area	Arrivals 1992	Foreign-currency income from tourism
	(1,000)	(1,000)	(%)	(sq.km.)	(per sq.km. of land area)	(%) 1992
France	59,590	42,721	+39	547,026	108.9	9.7
Austria	19,098	16,571	+15	83,856	227.7	25.0
Germany	15,147	14,501	+4	356,910	42.4	2.5
Switzerland	12,800	11,700	+9	41,293	310.0	10.4
Belgium	12,740[1]	10,576[1]	+20[1]	30,513	417.5[1]	3.2
Netherlands	6,049	4,876	+24	41,547	145.6	3.6
Luxemburg	796	760	+5	2,586	307.8	?
Monaco	246	232	+6	2	123.0	?
Liechtenstein	72	72	0	160	450.0	?

[1] unclear if daytime tourists included

Tourism statistics, Southern Europe

	Arrivals 1992	Arrivals 1988	Change	Land area	Arrivals 1992	Foreign-currency income from tourism
	(1,000)	(1,000)	(%)	(sq.km.)	(per sq.km. of land area)	(%) 1992
Spain	39,638	38,784	2	504,782	78.5	25.6
Italy	26,113	26,155	0	301,268	86.7	10.8
Greece	9,331	7,923	+18	132,000	70.7	25.6
Portugal	8,884	6,624	+34	91,985	96.6	17.2
Croatia	2,010	?	-84[1]	56,538	35.6	?
Malta	1,002	784	+28	316	3,170.9	26.9
Slovenia	616	?	?	20,256	30.4	10.8
San Marino	582[2]	504	+15	61	9,541.0[2]	?
Yugoslavia (rest)	156	1,290	-88	102,173	1.5	?
Macedonia	106	241	-56	25,713	4.1	?
Gibraltar	88	156	-44	6	1,466.7	?

[1] all tourists

[2] 1990

Tourism statistics, Eastern Mediterranean

	Arrivals 1992	Arrivals 1988	Change	Land area	Arrivals 1992	Foreign-currency income from tourism
	(1,000)	(1,000)	(%)	(sq.km.)	(per sq.km. of land area)	(%) 1992
Turkey	6,549	3,715	+76	779,452	8.4	19.6
Cyprus	1,991	1,112	+79	9,251	215.2	60.6
Israel	1,502	1,170	+28	20,770	72.3	12.5

Appendix F. Alps Convention Draft Protocol on Tourism

(Excerpt; Version of November 15, 1995; unofficial translation)

Chapter I. General Provisions

Article 1. Objective

The objective of this Protocol is to contribute to lasting development through sustainable tourism within the framework of the existing governmental structures, using binding measures and recommendations which are mindful of the interests of the native populace and visitors.

Article 2. International co-operation

(1) The Contracting Parties shall remove impediments to international co-operation between the regional bodies of the Alpine Region and further the solution of common problems at the most appropriate level.

(2) The Contracting Parties shall support an intensified international co-operation between the responsible institutions. In particular, they shall be mindful of enhancing the status of border areas by co-ordinating sustainable tourism and leisure-time activities.

(3) If the determination of measures is the responsibility of national or international authorities, the regional bodies shall be given the possibiliy of effectively stating the interests of the populace.

Article 3. Consideration of (touristic) objectives in other policies

The Contracting Parties shall respect the objectives of this Protocol in their other policies, as well, in particular in the fields of regional planning, (city planning), transportation, agriculture, forestry, environmental and natural protection and water and energy supply, (waste disposal), in order to lessen potential adverse effects or impacts which are contrary to these objectives.

Article 4. Participation of regional bodies

(1) Each Contracting Party shall determine within the framework of the existing governmental structures the most appropriate level for the co-ordination and co-operation between the directly affected institutions and regional bodies, in order to further common responsibility; in particular, to utilise and further develop synergetic energies in the execution of tourism policies and the implementation of the resulting measures.

(2) The directly affected regional bodies shall be involved in the various stages of preparation and implementation of these policies and measures while preserving their responsibility within the existing government structures.

Chapter II. Specific Measures

Article 5. Controlled development of tourism

(1) The Contracting Parties shall be mindful of developing a sustainable and environmentally compatible tourism. To this end they shall support the drawing up and implementation of guidelines, (development) programmes and sectoral plans which will be introduced by the responsible authorities at the most appropriate level and do justice to the objectives of this Protocol.

(2) The measures shall above all enable the advantages and disadvantages of the planned developments to be assessed and compared, especially in regard to the following aspects:

a) socio-economic impacts on the populace,
b) impacts on soil, water, air, ecology and scenery with regard to the specific ecological conditions, natural resources and the carrying capacity of the ecosystems,
c) impacts on public finances.

Article 6. Orientation of touristic development

(1) The Contracting Parties shall incorporate the concerns of nature conservation and landscape management in tourism promotion. They shall as far as possible only promote such projects which conserve the landscape and environment

(2) They shall pursue a policy geared to permanence which bolsters the competitiveness of tourism in natural surroundings in the Alpine Region and thus makes an important contribution to the socio-economic development of the Alpine Region. In this regard, measures shall be given preference which promote innovation and the qualitative improvement of the range of tourism options.

(3) The Contracting Parties shall be mindful that in the highly frequented (tourism) areas the envisaged objective is a balance between intensive and extensive forms of tourism.

(4) In promotional measures the following aspects shall be taken into account:

a) in intensive tourism, the adaptation of existing touristic structures and facilities to ecological requirements and the development of new structures in consonance with the objectives of this Protocol;
b) in extensive tourism, the preservation or development of sustainable forms of tourism in natural surroundings and the enhanced status of the natural and cultural heritage of the tourist areas.

Article 7. Promotion of quality

(1) The Contracting Parties shall pursue a policy which is constantly and consistently geared to securing high-quality tourism in the entire Alpine Region, with special regard to ecological requirements.

(2) They shall further the exchange of information and the carrying out of common action programmes with the aim of improving quality, particularly in the following fields:

a) adaptation of amenities and facilities to landscape and nature,
b) city planning and architecture (new buildings and city renovation),
c) accommodations and touristic amenities,
d) diversification of tourism within the Alpine Region by enhancing the status of cultural activities in the pertinent areas.

Article 8. Controlling the streams of visitors

The Contracting Parties shall support organisational and planning measures to control the streams of visitors, particularly in protected areas, in order to secure their further existence.

Article 9. Limiting development in (natural) space

The Contracting Parties shall be mindful that touristic development be adapted to the specific environmental peculiarities and the available resources of a place or region. Projects with potentially significant impacts on the environment shall be subjected to a prior assessment within the framework of existing procedures, whose results shall be taken into account in reaching a final decision.

Article 10. Zones free of touristic activities

The Contracting Parties shall designate zones as free of touristic activities in accordance with their regulations and ecological criteria.

Article 11. Policy in the accommodations sector

The Contracting Parties shall develop policies in the accommodations sector which, while mindful of the limited space available in accommodations available for rent, give preference to the renewal and utilisation of existing building fabric and the modernisation and quality improvement of existing accommodation facilities.

Article 12. Ski-lift facilities

(1) The Contracting Parties shall agree to pursue a policy within the framework of national licensing procedures for ski-lift facilities which, apart from the requirements of profitability and safety, also gives due consideration to ecological and landscape concerns.

(2) Operation permits and concessions for ski-lift facilities shall be required to provide for the dismantling and removal of no longer needed such facilities and the natural restoration of no longer utilised surfaces with indigenous vegetation.

Article 13. Transportation and tourism

The Contracting Parties shall take steps geared to limiting motorised traffic in the touristic centres.

In addition, they shall support private or public initiatives designed to improve the accessibility of tourist areas (places or centres) by public transportation and the use of public transportation by tourists.

Article 14. Special development techniques

(1) Ski runs

a) The Contracting Parties shall be mindful of constructing, maintaining and using ski runs and ski operations which (as far as possible) are non-detrimental to the landscape. Moreover, solutions shall be sought which give due consideration to natural cycles and the fragility of biotopes.
b) Manipulation(s) of the terrain shall be avoided as far as possible and, inasmuch as the given natural areas permit, the newly shaped surfaces shall be planted with indigenous plants.

(2) Artificial snow production

(Majority proposal)

The production of snow can be permitted to guarantee a minimum supply for local and regional skiing operations or to secure exposed zones, ecological, hydrological and climatic conditions permitting.

These facilities shall only be operated during the customary natural skiing season in the Alps, which is to be determined according the given conditions of the area.

Article 15. Sports activities

(1) The Contracting Parties shall (particularly in protected areas) establish a policy to control and, where needed, prevent sports activities in the open in order that the environment suffers no adverse effects from them. When necessary, bans shall also be declared.

(2) The Contracting Parties shall restrict motor sports (outside of public transportation routes) as far as possible or where necessary ban them except where certain zones have been designated by the responsible authorities for such activities.

Article 16. Skydiving

The Contracting Parties shall restrict as far as possible skydiving outside of airports for sports purposes and where necessary ban or severely restrict it.

Article 17. Development of economically depressed areas

The Contracting Parties shall examine at the most appropriate level co-ordinated solutions to ensure a balanced development in economically depressed areas.

Article 18. Staggering of tourist season

(1) The Contracting Parties shall endeavour to develop a better spatial and seasonal staggering of touristic demand in holiday areas.

(2) To this end the international co-operation in staggering the tourist season and exchange of information on possibilities of lengthening the season shall be supported.

Article 19. Innovation incentives

The Contracting Parties shall develop suitable incentives for the implementation of the concerns of this Protocol. To this end they shall in particular consider establishing a competition in the Alpine countries to award prizes for innovative touristic projects and products which are in accordance with the objectives of this Protocol.

Article 20. Co-operation between the tourist industry, agriculture, forestry and the trades

The Contracting Parties shall support the co-operation between the tourism industry, agriculture, forestry and the trades. In this regard they shall in particular support commercial groupings which are conducive to expanding the work force with respect to sustainable development.

Article 21. More stringent measures

The Contracting Parties can take steps in accordance with the objectives of this Protocol which exceed the scope of the steps provided in this Protocol.

Appendix G. Antarctic Treaty Protocol on Environmental Protection (excerpts)

Article 3. Environmental principles

(1) The protection of the Antarctic environment and dependent and associated ecosystems and the intrinsic value of Antarctica, including its wilderness and aesthetic values and its value as an area for the conduct of scientific research, in particular research essential to understanding the global environment, shall be fundamental considerations in the planning and conduct of all activities in the Antarctic Treaty area.

(2) To this end:

a) activities in the Antarctic Treaty area shall be planned and conducted so as to limit adverse impacts on the Antarctic environment and dependent and associated ecosystems;
b) activities in the Antarctic Treaty area shall be planned and conducted so as to avoid:

   I) adverse effects on climate and weather patterns;
   II) significant adverse effects on air or water quality;
   III) significant changes in the atmospheric, terrestrial (including aquatic), glacial or marine environments;
   IV) detrimental changes in the distribution, abundance or productivity of species or populations of species of fauna and flora;
   V) further jeopardy to endangered or threatened species or populations of such species; or
   VI) degradation of, or substantial risk to, areas of biological, scientific, historic, aesthetic or wilderness significance;

c) activities in the Antarctic Treaty area shall be planned and conducted on the basis of information sufficient to allow prior assessments of, and informed judgments about, their possible impacts on the Antarctic environment and dependent and associated ecosystems and on the value of Antarctica for the conduct of scientific research; such judgments shall take full account of:

I) the scope of the activity, including its area, duration and intensity;
II) the cumulative impacts of the activity, both by itself and in combination with other activities in the Antarctic Treaty area;
III) whether the activity will detrimentally affect any other activity in the Antarctic Treaty area;
IV) whether technology and procedures are available to provide for environmentally safe operations;
V) whether there exists the capacity to monitor key environmental parameters and ecosystem components so as to identify and provide early warning of any adverse effects of the activity and to provide for such modification of operating procedures as may be necessary in the light of the results of monitoring or increased knowledge of the Antarctic environment and dependent and associated ecosystems; and
VI) whether there exists the capacity to respond promptly and effectively to accidents, particularly those with potential environmental effects;

d) regular and effective monitoring shall take place to allow assessment of the impacts of ongoing activities, including the verification of predicted impacts;
e) regular and effective monitoring shall take place to facilitate early detection of the possible unforeseen effects of activities carried on both within and outside of the Antarctic Treaty area on the Antarctic environment and dependent and associated ecosystems.

(3) Activities shall be planned and conducted in the Antarctic Treaty area so as to accord priority to scientific research and to preserve the value of Antarctica as an area for the conduct of such research, including research essential to understanding the global environment.

(4) Activities undertaken in the Antarctic Treaty area pursuant to scientific research programmes, tourism and all other governmental and non-governmental activities in the Antarctic Treaty area for which advance notice is required in accordance with Article VII (5) of the Antarctic Treaty, including associated logistic support activities, shall:

a) take place in a manner consistent with the principles in this Article; and
b) be modified, suspended or cancelled if they result in or threaten to result in impacts upon the Antarctic environment or dependent and associated ecosystems inconsistent with those principles.

...

Article 8. Environmental-impact assessment

(1) Proposed activities referred to in paragraph 2 below shall be subject to the procedures set out in Annex I for prior assessment of the impacts of those activities on the Antarctic environment or on dependent and associated ecosystems according to whether those activities are identified as having:

a) less than a minor or transitory impact;
b) a minor or transitory impact; or
c) more than a minor or transitory impact.

(2) Each Party shall ensure that the assessment procedures set out in Annex I are applied in the planning processes leading to decisions about any activities undertaken in the Antarctic Treaty area pursuant to scientific research programmes, tourism and all other governmental and non-governmental activities in the Antarctic Treaty area for which advance notice is required under Article VII (5) of the Antarctic Treaty, including associated logistic support activities.

(3) The assessment procedures set out in Annex I shall apply to any change in an activity whether the change arises from an increase or decrease in the intensity of an existing activity, from the addition of an activity, the decommissioning of a facility, or otherwise.

(4) Where activities are planned jointly by more than one Party, the Parties involved shall nominate one of their number to co-ordinate the implementation of the environmental-impact-assessment procedures set out in Annex I.

Appendix H.
Guideline for Visitors in Antarctica
(excerpt)

...

This Guideline for visitors to Antarctica is intended to ensure that all visitors know the (Antarctic) Treaty and the (environment-protection) Protocol and can therefore fulfill their provisions. Visitors are of course bound by national laws and other regulations applicable to activities in Antarctica.

A) Protect the wild species in Antarctica

Removing anything from nature or behaving in a way which has a harmful effect on wild species of the Antarctic is prohibited unless permission has been granted by a national authority.

1. Do not use aircraft, ships, small boats or other means of transportation in a manner such that wild species of the sea or land are disturbed.

2. Do not feed or touch birds or seals. Approaching them or photographing them in a way which leads to a change in their behaviour is prohibited. Special precaution is in order with nesting or molting animals.

3. Do not damage plants, e.g., by trampling on them, by driving over or landing on extended moss cushions or on lichen-covered scree slopes.

4. Use no rifles or explosives. The generation of noise is to be restricted to the lowest possible level so that the wild species are not frightened.

5. Bringing non-indigenous plants or animals to the Antarctic (e.g., living poultry, dogs and cats, culture plants) is prohibited.

B) Respect protected Areas

A great number of areas in the Antarctic are under special protection because of their particular ecological, scientific, historic or other value. Entering certain areas can be banned unless it has been granted by a permit issued by a national authority. Activities in or near designated historic sites and monuments and in certain other areas can be subject to certain restrictions.

1. The location of areas under special protection and all restrictions as to entering these areas and the activities which may be undertaken in or near them must be known.
2. Observe the restrictions in force.
3. Do not damage, remove or disturb historic sites and monuments or artefacts connected with them.

C) Respect scientific research

Do not impede scientific research, scientific facilities or their equipment.

1. Before visiting the Antarctic research and logistical-support facilities a permit must be obtained; it is to be confirmed 24 to 72 hours before arrival; the instructions in force for such visits are to be heeded to the letter.
2. Do not change or remove scientific apparatuses or marking posts and do not cause confusion in places of scientific study, field camps or supplies.

D) Be aware of safety precautions

Be prepared for harsh and changeable weather. Your gear and clothing must be adapted to Antarctic conditions. Bear in mind that the Antarctic environment is inhospitable, unpredictable and potentially dangerous.

1. Make a correct assessment of your capabilities and the dangers inherent in the Antarctic environment and act accordingly. Always plan your activities with your own safety in mind.
2. Keep a safe distance away from all wild animals and plants, on water and on land.
3. Observe and heed the advice and instructions of the guides; do not leave your group.
4. Do not walk on glaciers or large snowfields without appropriate gear and experience; there is a great danger of falling into concealed crevasses.
5. Do not count on rescue services; through reasonable planning, high-quality gear and trained personnel the independence of outside help is improved and there is less risk.
6. Do not enter emergency shelters (except in emergencies). If you use gear or food from a shelter, the nearest research station or government installation must be immediately informed of the emergency.

7. Respect no-smoking areas, particularly around buildings, and take great care to prevent the danger of fire. In the dry environment of Antarctica fire is a great hazard.

E) Preserve the untouched state of Antarctica

Antarctica is relatively untouched and has never been subjected to large-scale disturbances by man. It is the greatest wilderness on earth. This should stay this way.

1. Do not dispose of waste and garbage on land. Open fire is forbidden.
2. Do not impair or pollute lakes and rivers. Substances which are to be disposed of at sea must be disposed of according to regulations.
3. Do not paint or scratch names or graffiti on stones and buildings.
4. Do not collect or remove biological or geological samples or man-made artefacts as souvenirs, including stones, bones, eggs, fossils and parts or contents of buildings.
5. Do not disfigure or destroy any buildings whether they are inhabited, abandoned or uninhabited or shelters.

Appendix I. Guideline for Those Organising and Carrying Out Tourism and Non-governmental Activities in the Antarctic (excerpt)

The Protocol concerning Protection of the Environment designates Antarctica as a nature reserve devoted to peace and science and is valid for governmental and non-governmental activities in the Antarctic Treaty area. The Protocol is designed to ensure that human activity, including tourism, has no adverse effects on the Antarctic environment nor its scientific and aesthetic values.

The Protocol determines that in principle all activities must be planned and conducted on the basis of sufficient information to permit an assessment of their possible impacts on the Antarctic environment and associated ecosystems and on the value of Antarctica for the conduct of scientific research. Tour organisers should be aware that the Protocol on environmental protection demands that "activities be modified, suspended or cancelled if they result in or threaten to result in impacts upon the Antarctic environment or dependent and associated ecosystems."

Those responsible for the organisation and carrying out of tourism and non-governmental activities in the Antarctic must follow the pertinent national laws and other regulations implementing the Antarctic Treaty system as well as other state laws and other regulations implementing the international agreements on environmental protection, prevention of pollution and security with respect to the Antarctic Treaty area. They are moreover to heed the regulations which are made mandatory for tour organisers and ship operators in accordance with the Protocol on environmental protection and its annexes insofar as they have not yet found application in national legislation.

The Most Important Obligations for Tour Organisers and Ship Operators

1. Notify the responsible authorities of the pertinent Contracting Party or Contracting Parties on your activities before they begin, and report on them.
2. Conduct an assessment of the possible impacts of your intended activity on the environment.

3. Take care to have effective countermeasures ready for emergencies which would threaten the environment, particularly in regard to marine pollution.
4. Ensure your non-dependence on outside help and the safety of your untertakings.
5. Respect scientific research and the Antarctic environment, including the restrictions in regard to protected areas and the protection of the animal and plant worlds.
6. Prevent the disposal and discharge of waste.

Procedures to Be Followed by Tour Organisers and Ship Operators

A) In planning travel in Antarctica tour organisers and ship operators shall

1. notify the responsible authorities of the pertinent contracting party or the contracting parties as to details of their intended activity in time for the contracting party (parties) to (be able to) fulfill their obligation to exchange information in accordance with Article VII, Section 5 of the Antarctic Treaty. The information to be provided is listed in Annex A;
2. conduct an environmental-impact assessment in compliance with the procedures stipulated by national legislation to implement Annex A of the Protocol, if need be including the designation of the manner of monitoring possible impacts;
3. obtain a permit in advance from the national authorities responsible for the stations which they intend to visit;
4. provide information on the support for working out the following plans: plans for countermeasures to be taken in emergencies in accordance with Article 15 of the Protocol; waste-treatment plans in accordance with Annex III of the Protocol and plans to be followed in combatting marine pollution in accordance with Annex IV of the Protocol;
5. ensure that leaders of expeditions and travellers are aware of the location and the particular order of the specially protected areas and the sites of particular scientific interest (and, after the Protocol is in force, the specially protected areas of the Antarctic and specially administered areas of the Antarctic), the historic sites and monuments and in particular the corresponding administrative plans;
6. obtain a permit from the responsible national authority of the pertinent contracting party or the contracting parties insofar as this is required by national law if they have reason to enter such areas or a monitoring site designated in accordance with the Convention on the Conservation of Antarctic Marine Living Resources (CCAMLR Ecosystem Monitoring Programme);
7. ensure that activities can be completely conducted without outside help and require no support from the Contracting Parties insofar as arrangements on this have not been made in advance ;
8. ensure that they employ experienced and trained personnel, including a sufficient number of guides;

9. see to it that equipment, vehicles, ships and aircraft are used which are suitable for use in the Antarctic;
10. are thoroughly familiar with the applicable telecommunications, navigation, air-traffic-control and emergency procedures;
11. procure the best available maps and hydrographic charts in the awareness that not all areas have been completely or precisely explored;
12. settle the question of insurance (subject to the requirements of national law);
13. work out and conduct information and training programmes in order to ensure that their entire staff and all visitors are instructed on the applicable provisions of the Antarctic Treaty system;
14. provide all visitors with a copy of the "Guideline for Visitors in Antarctica".

B) During their stay in the Antarctic Treaty area tour organisers and ship operators shall

1. follow all the requirements of the Antarctic Treaty system and the pertinent national laws and ensure that visitors are instructed on the requirements applicable to them;
2. confirm arrangements to visit research stations 24 to 72 hours prior to their arrival and ensure that visitors are instructed on all conditions and restrictions laid down by the station;
3. ensure that visitors are supervised by a sufficiently large number of guides who have sufficient experience and training regarding the conditions in the Antarctic and knowledge of the requirements of the Antarctic Treaty system;
4. monitor the effects of their activity on the environment, if need be, and notify the responsible national authorities of the pertinent contracting party or the contracting parties about all adverse and cumulative impacts of an activity which were not foreseeable by the environmental-impact assessment;
5. operate ships, yachts, small boats, aircraft, hovercraft and all other means of transportation safely and in compliance with the appropriate procedures, including those listed in the Antarctic Flight Information Manual (AFIM);
6. dispose of waste substances in compliance with Annexes III and IV of the Protocol. These annexes prohibit, among other things, the discharge of plastic objects, oil and harmful substances in the Antarctic Treaty area, regulate the discharge of sewage and food waste and require the removal of most waste products from the area;
7. fully co-operate with the observers named by the consultative parties to carry out inspections of stations, ships, aircraft and equipment in accordance with Article VII of the Antarctic Treaty and with the observers to be named in accordance with Article 14 of the environmental-protection protocol;
8. participate in monitoring programmes in accordance with Article 3, Section 2, Littera (d) of the Protocol;
9. draw up a meticulous and complete report on the activities conducted.

C) After completion of the activities

Within three months after completion of the activity, tour organisers and ship operators shall submit a report on this activity to the responsible national authority in accordance with national laws and procedures. These reports should contain the following information: the name of and information on each ship or aircraft used, its country of registration and the name of its captain or commander; the itinerary actually taken; the number of visitors participating in the activity; places, dates and purposes of landing and the number of visitors who went ashore each time; all meteorological observations including those which were made as part of the Voluntary Observing Ships Scheme of the World Meteorological Organisation (WMO); all important changes of the activities and their impacts as compared to the impacts predicted before the travel was carried out; and the activities carried out in emergencies.

Appendix J. Council of Europe
Recommendation No. R (94) 7

of the Committee of Ministers to Member States

On a General Policy for Sustainable and Environment-Friendly Tourism Development

The Committee of Ministers, under the terms of Article 15 (b) of the Statute of the Council of Europe;

Considering that the aim of the organisation is to achieve a greater unity between its members, inter alia in order to foster their economic and social progress;

Having regard to the various activities carried out within the Council of Europe and other international organisations;

Bearing in mind the declaration of the ministerial conference held in Lucerne from 28 to 30 April 1993 on "Environment for Europe", which calls on the Council of Europe to pursue its activities to promote ecologically viable tourism;

Taking into account the declaration of the United Nations Conference on Environment and Development held in Rio from 3 to 14 June 1992;

Stressing that tourism constitutes one of the mainsprings of economic growth and is likely to become the foremost world industry;

Acknowledging that tourism is a factor in bringing people together, forging a European identity and heightening awareness of the value of their natural and cultural heritage;

Noting a growing interest in all forms of tourism associated with the discovery and knowledge of the natural and cultural heritage;

Convinced that the environment has an intrinsic value which is greater than its value

as a tourism asset;
Underlining that the relationship beween tourism and the environment is a delicate one;

Aware of the threats posed to the natural and landscape environment and local populations and cultures by the excessive and uncontrolled development of tourism;

Observing that levels of tourism development and fragility of the areas concerned vary from one country to another, and even from one region to another;

Convinced of the need to establish a general framework in order to safeguard and restore the quality of the environment, which is the prime resource of tourism.

Recommends that the governments of member states:

a. base their tourism development policy on the principles and measures set out in the appendix to this recommendation, tailoring them where necessary to the special features or fragility of certain regions;
b. ensure that the national, regional and local authorities, those institutions responsible for tourism and the environment, the tourist industry and all other sectors involved be duly informed of this recommendation and respect the principles contained herein.

Instructs the Secretary General to convey this recommendation to the international organisations and international financial bodies working in the field of the development of tourism.

Appendix to Recommendation No. R (94) 7

I. General Principles

1. The principles of prevention, precaution and remedial action allied with the need for sustainable development should underlie any tourism development policy.

2. Every planned tourism activity or development should be geared to sustainable development and its impact on the environment should be assessed; environmental considerations should be integrated into the decision-making process from the start of the project.

3. In principle, no permission should be given for any project having a significant environmental impact without evidence of its environmental, economic and financial viability. However, if this cannot be proven, other considerations may be taken into account, such as the project's contribution to socio-cultural development.

4. Tourism development should be totally or partly self-financing where possible, with the emphasis on achieving or building on sustainability.

5. Tourism development should be a gradual process and not outstrip infrastructure improvements. Tourism projects must be carried out within the limits of the local infrastructure.

6. Tourism should be developed so that in addition it benefits the local community, provides support for the local economy and takes account of the latter's ability to absorb development. Wherever possible, it should encourage employment of the local workforce and use local materials and traditional skills.

7. Tourism activities and amenities should be located in carefully chosen areas so as to restrict development in sensitive regions. Wherever possible, before building any new amenities, the possibilty of using, modernising or rehabilitating existing infrastructures should first be considered.

8. Tourism activities and development must respect the scale, nature, character and capacity of the local physical and social environment of the place in which they are sited, as well as its natural resources, landscape quality, historic and archaeological heritage and cultural identity.

To this end, every project should be subjected to an environmental impact assessment. Where an environmental impact assessment (EIA) is required, due to the nature and size of the project and the character of the area to be affected, the following elements should be included:

- the impact on environment, landscape, fauna, flora, water, land and energy resources;
- the impact on local infrastructure, economy, society and employment;
- the direct, indirect, immediate and long-term effects of the project;
- the effects of secondary developments (transport, new infrastructures, etc.);
- adequate consultation with the local public and local communities;
- possible remedial or compensating measures.

9. Therefore, where appropriate, every tourism project should:

- avoid creating additional pressures on the environment;
- encourage the use of public and non-motorised transport, as well as the most suitable technology for saving water and energy, treating effluent and processing and recycling waste;
- be accompanied by a monitoring programme to ensure that once in operation, the project keeps to its environmental commitments, and that unforeseen negative impacts are detected and dealt with immediately. Eco-audits should be carried out on large-scale developments;
- strive to make visitors aware of the need to protect the environment and the constraints that this protection entails.

II. Implementation of the General Principles

The implementation of the general principles will be at international, national, regional and local level.

A. National authorities

1. In order to implement the general principles national authorities should:

a. develop national strategies for sustainable and environmentally friendly development;
b. ensure that the various policies and decision-making levels are coherent and consistent. Tourism development is to be considered in terms of both the overall economy of a country and the local economy;
c. attempt to draw up an inventory of the country's cultural and natural resources and set up a legislative framework to enhance and protect them where necessary;
d. preserve areas designated as vulnerable by following a policy of land use control and through purchase, renting and management agreements;
e. draw up a framework for integrated planning and resource management;
f. draw up a national tourism policy taking full account of the environment and defining the role and importance of tourism in the national economy; such a policy should be tied in with overall planning policy;
g. ensure close collaboration between bodies responsible for providing reliable statistics on the tourist industry and monitoring the state of the country's environment;
h. provide environmental education and training for tourism professionals and ensure that training in the tourism sector builds awareness of the environment;
i. run campaigns to build awareness of the environment among local communities, elected representatives in tourist destinations and tourists themselves;
j. propose charters for the tourist industry establishing qualitative criteria for environmentally friendly tourism;
k. control tourist demand and flow, in particular by staggering the tourist season and easing pressure on certain sites by developing other centres of interest, introducing admission fees at certain sites or for certain services, or limiting the number of tourists;
l. ensure that sites harmed by excessive tourism are restored as appropriate;
m. diversify what is on offer for tourists by encouraging new types of activities as alternatives to mass tourism, based on an interest in the country concerned and knowledge of its heritage, culture and way of life;
n. encourage the introduction of environmentally sound products and activities by relevant measures including the awarding of prizes and quality labels;
o. consider, where appropriate, the development of a tax incentive scheme to encourage environment-friendly tourism development projects;
p. consider the possibility of introducing sanctions penalising those responsible for activities harmful to the environment, geared above all to preventing harm; these should include sufficient levels of compensation to be paid by developers for the repair of any damage caused during the development phase.

2. National authorities, acting through international organisations, should propose that these organisations:

a. adopt an integrated planning approach to future development of the tourism industry, emphasising the need to protect the social, natural and cultural environments;
b. promote international awards for sustainable tourism respecting the environment;
c. publish international guides of good tourism practice vis-à-vis the environment, including databases of relevant documents and successful projects;
d. support training and awareness-building programmes on tourism and the environment;
e. support pilot projects for sustainable tourism and disseminate information about them.

3. National authorities, acting through international financial bodies, should propose that these bodies:

a. request environmental impact statements for all projects they finance and run impact assessment studies (EIA) themselves where appropriate;
b. ensure that all relevant procedures have been complied with;
c. ascertain the environmental viability of the project on the same footing as the economic and financial viability of the project;
d. make suitable supervision arrangements to check that the project is properly run;
e. encourage the use of the most appropriate technology so as to minimise the impact on the environment.

B. Local and regional authorities

In order to implement the general principles local and regional authorities should:

a. exercise control over tourist development with potentially significant environmental consequences through regional, local and urban planning policy and a policy of nature and landscape protection;
b. establish local tourism plans, based on inventories of sites and biotopes and on their carrying capacity and social accommodation capacity as well as land use plans; local tourism plans should be integrated into overall local development plans, and local tourism development funding should be provided in the framework of these plans where appropriate;
c. make the issue of building permits contingent on requirements such as a guarantee of quality development and respect for the environment, and ensure that these criteria are respected;
d. work closely with all public and private-sector operators to ensure co-ordination between different tourism development projects, and maintain a regular, two-way flow of information.

Appendix K. Charter for Sustainable Tourism

We, the participants at the World Conference on Sustainable Tourism, meeting in Lanzarote, Canary Islands, Spain, on 27–28 April 1995,

Mindful that tourism, as a worldwide phenomenon, touches the highest and deepest aspirations of all people and is also an important element of socioeconomic and political development in many countries,

Recognizing that tourism is ambivalent, since it can contribute positively to socioeconomic and cultural achievement, while at the same time it can contribute to the degradation of the environment and the loss of local identity, and should therefore be approached with a global methodology,

Mindful that the resources on which tourism is based are fragile and that there is a growing demand for improved environmental quality,

Recognizing that tourism affords the opportunity to travel and to know other cultures, and that the development of tourism can help promote closer ties and peace among peoples, creating a conscience that is respectful of the diversity of culture and life styles,

Recalling the Universal Declaration of Human Rights, adopted by the General Assembly of the United Nations, and the various United Nations declarations and regional conventions on tourism, the environment, the conservation of cultural heritage and on sustainable development,

Guided by the principles set forth in the Rio Declaration on the Environment and Development and the recommendations arising from Agenda 21,

Recalling previous declarations on tourism, such as the Manila Declaration on World Tourism, the Hague Declaration and the Tourism Bill of Rights and Tourist Code,

Recognizing the need to develop a tourism that meets economic expectations and environmental requirements, and respects not only the social and physical structure of destinations, but also the local population,

Considering it a priority to protect and re-enforce the human dignity of both local communities and tourists,

Mindful of the need to establish effective alliances among the principal actors in the field of tourism so as to fulfil the hope of a tourism that is more responsible towards our common heritage,

APPEAL to the international community and, in particular, URGE governments, other public authorities, decision-makers and professionals in the field of tourism, public and private associations and institutions whose activities are related to tourism, and tourists themselves, to adopt the principles and objectives of the Declaration that follows:

1. Tourism development shall be based on criteria of sustainability, which means that it must be ecologically bearable in the long term, as well as economically viable, and ethically and socially equitable for local communities.

Sustainable development is a guided process which envisages global management of resources so as to ensure their viability, thus enabling our natural and cultural capital, including protected areas, to be preserved. As a powerful instrument of development, tourism can and should participate actively in the sustainable development strategy. A requirement of sound management of tourism is that the sustainability of the resources on which it depends must be guaranteed.

2. Tourism should contribute to sustainable development and be integrated with the natural, cultural and human environment; it must respect the fragile balances that characterize many tourist destinations, in particular small islands and environmentally sensitive areas. Tourism should ensure an acceptable evolution as regards its influence on natural resources, biodiversity and the capacity for assimilation of any impacts and residues produced.

3. Tourism must consider its effects on the cultural heritage and traditional elements, activities and dynamics of each local community. Recognition of these local factors and support for the identity, culture and interests of the local community must at all times play a central role in the formulation of tourism strategies, particularly in developing countries.

4. The active contribution of tourism to sustainable development necessarily presupposes the solidarity, mutual respect and participation of all actors, both public and private, implicated in the process, and must be based on efficient cooperation mechanisms at all levels: local, national, regional and international.

5. The conservation, protection and appreciation of the worth of the natural and cultural heritage afford a privileged area for cooperation. This approach implies that all those responsible must take upon themselves a true challenge, that of cultural, technological and professional innovation, and must also undertake a major effort to create and implement integrated planning and management instruments.

6. Quality criteria both for the preservation of the tourist destination and for the capacity to satisfy tourists, determined jointly with local communities and informed by the principles of sustainable development, should represent priority objectives in the formulation of tourism strategies and projects.

7. To participate in sustainable development, tourism must be based on the diversity of opportunities offered by the local economy. It should be fully integrated into and contribute positively to local economic development.

8. All options for tourism development must serve effectively to improve the quality of life of all people and must influence the socio-cultural enrichment of each destination.

9. Governments and the competent authorities, with the participation of NGOs and local communities, shall undertake actions aimed at integrating the planning of tourism as a contribution to sustainable development.

10. In recognition of economic and social cohesion among the people of the world as a fundamental principle of sustainable development, it is urgent that measures be promoted to permit a more equitable distribution of the benefits and burdens of tourism. This implies a change of consumption patterns and the introduction of pricing methods which allow environmental costs to be internalised.

Governments and multilateral organizations should prioritize and strengthen direct and indirect aid to tourism projects which contribute to improving the quality of the environment. Within this context, it is necessary to explore thoroughly the application of internationally harmonized economic, legal and fiscal instruments to ensure the sustainable use of resources in tourism.

11. Environmentally and culturally vulnerable spaces, both now and in the future, shall be given special priority in the matter of technical co-operation and financial aid for sustainable tourism development. Similarly, special treatment should be given to zones that have been degraded by obsolete and high impact tourism models.

12. The promotion of alternative forms of tourism that are compatible with the principles of sustainable development, together with the encouragement of diversification, represent a guarantee of stability in the medium and the long term. In this respect there is a need, for many small islands and environmentally sensitive areas in particular, to actively pursue and strengthen regional co-operation.

13. Governments, industry, authorities, and tourism-related NGOs should promote and participate in the creation of open networks for research, dissemination of information and transfer of appropriate knowledge on tourism and environmentally sustainable tourism technologies.

14. The establishment of a sustainable tourism policy necessarily requires the support and promotion of environmentally compatible tourism management systems, feasibility studies for the transformation of the sector, as well as the implementation of demonstration projects and the development of international co-operation programmes.

15. The travel industry, together with bodies and NGOs whose activities are related to tourism, shall draw up specific frameworks for positive and preventive actions to secure sustainable tourism development and establish programmes to support the implementation of such practices. They shall monitor achievements, report on results and exchange their experiences.

16. Particular attention should be paid to the role and the environmental repercussions of transport in tourism, and to the development of economic instruments designed to reduce the use of non-renewable energy and to encourage recycling and minimization of residues in resorts.

17. The adoption and implementation of codes of conduct conducive to sustainability by the principal actors involved in tourism, particularly industry, are fundamental if tourism is to be sustainable. Such codes can be effective instruments for the development of responsible tourism activities.

18. All necessary measures should be implemented in order to inform and promote awareness among all parties involved in the tourism industry, at local, national, regional and international level, with regard to the contents and the objectives of the Lanzarote Conference.

Final Resolution

The World Conference on Sustainable Tourism considers it vital to make the following statements:

1. The Conference recommends State and regional governments to draw up urgently plans of action for sustainable development applied to tourism, in consonance with the principles set out in this Charter.

2. The Conference agrees to refer the Charter for Sustainable Tourism to the Secretary-General of the United Nations, so that it may be taken up by the bodies and agencies of the United Nations system, as well as by international organizations which have co-operation agreements with the United Nations, for submission to the General Assembly.

Bibliography

ABC, Aug. 14, 1994: España, Centro del Turismo Mundial. (Spanish newspaper). Madrid.

THE ADVENTURE TRAVEL SOCIETY 1991: Proceedings of the 1991 World Congress on Adventure Travel and Eco-Tourism, Colorado Springs, USA, Aug. 28–31.

THE ADVENTURE TRAVEL SOCIETY 1992: Proceedings of the 1992 World Congress on Adventure Travel and Eco-Tourism, Whistler, B.C., Canada, Sept. 20–23.

AGARDY, M.T. 1993: Accommodating Ecotourism in Multiple Use Planning of Coastal and Marine Protected Areas. In: Ocean & Coastal Management, vol. 20, 1993 (special issue: Ecotourism in Marine and Coastal Areas).

AGÖT 1995: Ökotourismus als Instrument des Naturschutzes? – Möglichkeiten zur Erhöhung der Attraktivität von Naturschutzvorhaben. Forschungsberichte des Bundesministeriums für wirtschaftliche Zusammenarbeit und Entwicklung [Ecotourism, an Instrument of Conservation?: Possibilities of Increasing the Attractiveness of Conservation Projects. Research Reports of BMZ], Weltforum Verlag, Cologne.

AKEGHEJO-SAMSONS, Y. 1993: Understanding Nigeria's Coastal Wetlands – MAB/UNESCO Theme 5 Project. In: Ocean & Coastal Management 19 (1993), Recent Developments and Announcements, Essex, U.K.

ANDERSEN, D.L., 1993: A Window to the Natural World – The Design of Ecotourism Facilities. In: THE ECOTOURISM SOCIETY 1993b.

ANDRESEN, F.H. 1993: Eine unendliche Geschichte? Befahrensregelung in den Wattenmeer-Nationalparken [A Never-Ending Tale? – A Regulation on Travel in the Wadden Sea National Park]. In: Nationalpark 4/93.

ANONYMUS 1995: La junta formenterá el desarollo local en los parcos naturales [The Commission Will Regulate Local Development in the Nature Parks]. In: Medio Ambiente, no. 23, April, 1995, Junta Andalusia Consejeria de Medio Ambiente, p. 14f.

BACHMANN, P. 1987: Tourism in Kenya – A Basic Need for Whom? Europäische Hochschulschriften, series X, Fremdenverkehr [Tourism], vol. 10, Bern.

BAILEY, C. 1991: Conservation and Development in the Galápagos Islands. In: WEST/BRECHIN 1991, part 5: Nature Preservation and Ecodevelopment: Ecotourism.

BEAZLEY, M. 1991: Oceans: A World Conservation Atlas. Mohndruck GmbH, Gütersloh, Germany.

BECHMANN A./FAHRENHORST, B. et al. 1993a: Entwicklung eines Modells für den Transfer ökologischen Know-hows aus Industrieländern in Entwicklungsländern. Umweltpolitik, Umweltplanung und Umweltverträglichkeitsprüfung in der Türkei. Ergebnisbericht [Development of a Model for the Transfer of Ecological Know-how from Industrial Countries to Developing Countries: Environmental Policy, Environmental Planning and Environmental-Impact Assessment in Turkey, Report of Findings], vol. IV, Berlin.

BECHMANN A./FAHRENHORST, B. et al. 1993b: Umweltverträglichkeitsprüfung im Süd-Antalya-Tourismusentwicklungsgebiet, Ergebnisbericht [Environmental-Impact Assessment in the South Antalya Tourism-Development Area, Report of Findings], vol. VIII, Berlin.

BECK, P.J. 1994: Managing Antarctic Tourism – Front-Burner Issue. In: Annals of Tourism Research, vol. 21, no. 2, 1994.

BECKER, P. 1994: Kahlschlag im tropischen Brackwasser – An den Küsten werden im großen Stil Mangrovenwälder vernichtet [Clear-Cutting in Tropical Brackish Waters: On the Coasts Mangrove Forests Are Being Destroyed Big-Time]. In: Der Tagesspiegel, May 25, 1994, Berlin.

BENTLEY, R. 1993: The Importance of the Environment for Tourism: Global Overview and Prospects. Presentation to the III International Symposium on Tourism, Ecology, and Municipalities, Aug. 30 to Sept. 4, 1993, in Mazatlán, México.

BIBBY et al. 1992: Putting Biodiversity on the Map – Priority Areas for Global Conservation. International Council for Bird Preservation, Cambridge.

BLAB, J./KLEIN, M./SSYMANK, A. 1995: Biodiversität und ihre Bedeutung in der Naturschutzarbeit [Biodiversity and Its Significance in Conservation Work]. In: Natur und Landschaft, vol. 70 (1995), no.1, Stuttgart.

BLUE PLAN 1989, see GRENON & BATISSE.

BOO, E. 1990: Ecotourism: The Potentials and Pitfalls, vol. 1. WWF, Washington, D.C.

BOO, E. 1990: Ecotourism: The Potentials and Pitfalls, vol. 2: Country Case Studies. WWF, Washington, D.C.

BRINKMANN, R./HEINS, J.U./KÖHLER, B. 1990: Menderes-Delta, Zustand und Gefährdung eines ostmediterranen Flußdeltas, Gemeinschaftsprojekt der TU Hannover und der GH Kassel [Menderes Delta: State and Threat to an Eastern Mediterranean River Delta. Joint Project of the Technological University of Hannover and the University of Kassel], Hannover.

BROUFMANN, A. 1993: Anthropogene Zerstörung des Seeufer-Ökosystems des Azowschen und Schwarzen Meeres. Ursachen. Entwicklungen. Lösungen [Anthropogenic Destruction of the Coastal Ecosystem of the Sea of Azov and the Black Sea: Causes, Developments, Solutions]. Beiträge der Akademie für Natur- und Umweltschutz Baden-Württemberg, vol. 16, Stuttgart.

BROUFMANN, A. 1993: The Environment in Eastern Europe – Out of the Communist Frying Plan into the Market Economy Fire. Proceedings of the International Colloquium, Stuttgart, Vienna.

BRÜGGEMANN, J. 1993: Auf der Suche nach dem grünen Paradies – Tourismus und Naturschutz in Costa Rica [In Search of Green Paradise: Tourism and Conservation in Costa Rica]. In: HÄUSLER et al. 1993.

BUFF, J. 1991: Wo Spanien wieder Waldland ist. Spanien-Exkursion des Nordwestdeutschen Forstvereins [Where Spain Is Once Again Forest Land: Spain Excursion of the Northwest German Forestry Association]. In: Forst und Holz, no.18, p. 505–508.

BUNDESAMT FÜR NATURSCHUTZ (BfN) (ed.) 1995: Materialien zur Situation der biologischen Vielfalt in Deutschland [Material on the Situation of Biological Diversity in Germany]. Bonn.

BUNDESMINISTERIUM FÜR UMWELT, NATURSCHUTZ UND REAKTORSICHERHEIT (BMU) 1992: Dokumente. Übersetzung der Agenda 21 [Documents. Translation of Agenda 21]. Bonn.

BMU (ed.) 1992: Konferenz der Vereinten Nationen für Umwelt und Entwicklung im Juni 1992 in Rio de Janeiro. Dokumente: Klimakonvention, Konvention über die Biologische Vielfalt, Rio-Deklaration, Walderklärung [United Nations Conference for Environment and Development in June, 1992 in Rio de Janeiro, Documents: Climate Convention, Convention on Biological Diversity, Rio Declaration, Forest Declaration]. Umweltpolitik series, Bonn.

BMU (ed.) 1993: Konferenz der Vereinten Nationen für Umwelt und Entwicklung im Juni 1992 in Rio de Janeiro. Dokumente: Agenda 21 [United Nations Conference for Environment and Development, Documents: Agenda 21]. Umweltpolitik series, Bonn.

BMU 1995: Umwelt No. 10/ 1994, Nos.2 and 10, 11, 12/1995, Eine Information des Bundesumweltministeriums [Environment, Information from Federal Ministry of the Environment], Bonn.

BUNDESMINISTERIUM FÜR WIRTSCHAFTLICHE ZUSAMMENARBEIT UND ENTWICKLUNG (BMZ) (ed.) 1993: Tourismus in Entwicklungsländern [Tourism in Developing Countries]. Compiled by: Aderhold, P., von Laßberg, D., Stäbler, M. and Vielhaber, A. Entwicklungspolitik series, Materialien, no. 88, Bonn.

BÜRO FÜR TOURISMUS- UND ERHOLUNGSPLANUNG (BTE) 1994: Beitrag zur Entwicklung eines anwendungsreifen Kriterienkataloges für ein Umweltgütesiegel in Tourismusgemeinden. Gutachten im Auftrag des Deutschen Fremdenverkehrsverbandes (DFV) [Contribution to the Development of an Applicable Catalogue of Criteria for Environmental-Quality Seals in Tourist Communities. Expert Report for the German Tourism Association], Bonn, Hannover (unpublished).

BTE/Planungsbüro WIRZ 1994: Landschaftsplanung und Fremdenverkehrsplanung [Landscape Planning and Tourism Planning], in: Bundesamt für Naturschutz (ed.) Angewandte Landschaftsökologie no. 1, Bonn-Bad Godesberg.

BTE 1995: Arbeitsmaterialien für einen umweltschonenden Tourismus. Campingplätze [Work Material for Sustainable Tourism: Camping Grounds], Berlin.

BURHENNE, W. n.d.: Internationales Umweltrecht – Multilaterale Verträge [International Environmental Law – Multilateral Treaties], loose-leaf edition, Berlin.

BURHENNE-GUILMIN, F. 1993: The Convention on Biological Diversity, in: Yearbook of International Environmental Law, vol. 3 (1992), 1993, p. 43 ff.

BURHENNE-GUILMIN, F./GLOWKA, L. 1994: An Introduction to the Convention on Biological Diversity, in: Widening Perspectives on Biodiversity, p. 15 ff., Gland/Geneva.

BUSH, P. 1993: The Cayman Islands – A Case Study for the Establishment of Marine Conservation Legislation in Small Island Countries. In: CTO 1993.

CAMPBELL, D. 1994: Ecotourism Booms in South-East Asia. In: The Ecotourism Society Newsletter, vol. 4, no. 4, fall 1994.

CARTER et al. 1993: Man's impact on the Coast of Ireland. In: WONG et al. 1993. Dordrecht.

CCE (Comisión de la Comunidad Europea, El Proyecto de Asistencia Técnica) / ICT (Instituto Costarricense de Turismo) 1993: Plan Estratégico de Desarrollo Turístico Sustentable de Costa Rica [Strategic Plan for Sustainable Touristic Development in Cosa Rica] (1993–1998), Síntesis Ejecutiva, San José, Costa Rica.

CDPE (Steering Committee for the Protection and Management of the Environment and Natural Habitats) 1995: Pan-European Biological and Landscape Diversity Strategy. Strasbourg.

CDT (Commonwealth Department of Tourism) 1994: National Ecotourism Strategy, Canberra.

CHAPE, S.P. 1990: Coastal Tourism and Environmental Management in Fiji. In: MILLER/AUYONG 1991.

CHUNG, C. 1994: Integrated Coastal Zone Management: A Policy Framework. In: Colloquy on the Protection of the Coastal Areas of the Adriatic Sea. Strasbourg.

COCKERELL, N. 1994: The Changing Role of Governments in Tourism. Press Release, ITB Berlin 1994.

COLWELL, S. 1995: personal communication, Jan. 5, 1995.

COMMISSION OF THE EUROPEAN COMMUNITIES 1992: Towards Sustainability: A European Community Programme of Policy and Action in Relation to the Environment and Sustainable Development. COM (92) 23 final – vol. II, Brussels, 27 March 1992.

CONSERVATOIRE DE L'ESPACE DU LITTORAL 1995: Atlas des espaces naturels du littoral [Atlas of Coastal Nature Areas], no. 68. Paris.

CORLAY, J.P. 1993: Coastal Wetlands – A Geographical Analysis and Some Projects for Management. In: Ocean & Coastal Management 19 (1993), Essex, U.K.

CTO (Caribbean Tourism Organization) 1991: First Caribbean Conference on Ecotourism, July 9–12, 1991, Belize City (proceedings).

CTO 1992: Tourism in Partnership with Nature – Second Caribbean Conference on Ecotourism, May 20–22, 1992, St. John, USVI (proceedings).

CTO 1993: Third Caribbean Conference on Ecotourism – Protecting the Caribbean Sea: Our Heritage, Our Future. May 4–7, 1993, Cayman Islands (proceedings compiled and edited by V. A. Brereton).

DANZ, W./ORTNER, S. (eds.) 1993: Die Alpenkonvention [The Alps Convention], Munich.

DAVIS, S.D./HEYWOOD, V.H./HAMILTON, A.C. 1994: Centers of Plant Diversity. A Guide and Strategy for Their Conservation. Vol. 1, published by The World Wide Fund for Nature (WWF) and IUCN – International Union for Conservation of Nature.

De KLEMM, C. 1993: Biological Diversity, Conservation and the Law, Gland/ Cambridge (U.K.).

DENZLER, S. 1993: Ansätze schweizerischer Reiseveranstalter für einen umweltverträglichen Drittwelttourismus [Approaches of Swiss Tour Operators for Sustainable Third-World Tourism]. In: Tagungsband zur Envirotour Vienna 1993, publ. by Internationale Gesellschaft für Umweltschutz, Vienna.

DER SPIEGEL 19/1995: Nur zur Probe. Wassernot auf der Urlaubsinsel Mallorca – Politiker schlagen Alarm, Umweltschützer protestieren [Just a Tryout: Water Shortage on the Holiday Island of Mallorca: Politicians Sound the Alarm, Environmentalists Protest], p.170f. Hamburg.

DEUTSCHE GESELLSCHAFT FÜR FREIZEIT (DGF) 1986: Freizeit-Lexikon [Leisure-Time Lexicon], Ostfildern.

DEUTSCHE GESELLSCHAFT FÜR FREIZEIT (DGF) 1995: Freizeit in Deutschland 1994/1995 [Leisure Time in Germany], Erkarth.

DEUTSCHER RAT FÜR LANDESPFLEGE [German Council for Land Management] 1984: Landespflege in Frankreich [Land Management in France], Schriftenreihe, no. 44.

DEUTSCHER REISEBÜROVERBAND e.V. (DRV) 1995: Umweltempfehlungen für touristische Zielgebiete [Environmental Recommendations for Tourist Destinations]. 3 nos., Frankfurt.

DEUTSCHES FREMDENVERKEHRSPRÄSIDIUM (ed.) 1994: Deutscher Tourismus-Bericht [German Tourism Report], Bonn/Frankfurt.

DIE TAGESZEITUNG (TAZ), Oct. 2, 1993: Öko-Touristen stören die Natur [Ecotourists Disrupt Nature], Berlin.

DIETRICH, K./KOEPFF, C. 1986: Erholungsnutzung des Wattenmeeres als Störfaktor für Seehunde [Recreational Use of the Wadden Sea, a Disturbance to Seals]. In: Natur und Landschaft, vol. 61 (1986), no. 7/8, Stuttgart

DIETRICH, K./KOEPFF, C. 1986: Wassersport im Wattenmeer als Störfaktor für brütende und rastende Vögel [Aquatic Sports in the Wadden Sea, a Disruption to Nesting and Resting Birds]. In: Natur und Landschaft, vol. 61, (1986), no. 6, Stuttgart.

DOMSCH, M., 1995: Umweltorientiertes Tourismusmanagement, dargestellt am Beispiel des Iberotels Sarigerme Park/Türkei [Environmentally Oriented Tourism Management: the Example of the Iberotel Sarigerme Park, Turkey]. Thesis, Dept. of Regional Recreation and Tourism, T.U. Berlin.

DOUGLAS, J.E. 1992: Ecotourism – The Future for the Caribbean? In: UNEP 1992.

DOWLING, R. 1993: An Environmentally-Based Planning Model for Regional Tourism Development. In: Journal of Sustainable Tourism, vol. 1, no. 1.

DROSTE, B. v./SILK, D./RÖSSLER, M. 1992: Tourism, World Heritage and Sustainable Development. In: UNEP 1992.
DUGAN, P. 1993: Wetlands in Danger – a MITCHELL BEAZLEY World Conservation Atlas. Mitchell Beazley, London.

ECOLETTER 1995: Sauberer Urlaub? [Clean Holiday?] no. 5/6, Munich.
ECTWT (Ecumenical Coalition on Third World Tourism) 1990: The Challenge of Tourism. Learning Resources for Study and Action.
EL PAIS, June 27, 1993 : Panamá y España convertirán el penal de Coiba en una isla para el ecoturismo [Panama and Spain Will Convert Coiba Prison into an Island for Ecotourism].
EL UNIVERSO, May 3, 1994: Ministro de Turismo: Es prioritario preservar Galápagos [The Minister of Tourism: It Is a Priority to Preserve the Galápagos], Guayaquil.
ELDER, D./PERNETTA, J. 1991: Oceans – a MITCHELL BEAZLEY World Conservation Atlas. Mitchell Beazley, London.
ELVERT, J. 1994: Irland mit Nordirland Reisehandbuch [Ireland and Northern Ireland: Travel Guide]. Stein Verlag, Kiel.
ENVIROTOUR VIENNA 1992: Strategies for Reducing the Environmental Impact of Tourism. Ed. by W. Pillmann and S. Predl. International Society for Environmental Protection, Vienna.
EPLER WOOD, M. 1994: International Update – The Green Globe Program. In: TES Newsletter, fall, 1994, vol. 4, no. 4.
ERDMANN, K.-H./NAUBER, J. 1993: Der deutsche Beitrag zum UNESCO-Programm "Der Mensch und die Biosphäre" (MAB) im Zeitraum Juli 1990 bis Juni 1992 [The German Contribution to the UNESCO Programme "Man and the Biosphere" from July, 1990 to June, 1992]. Rheinischer Landwirtschafts-Verlag, Bonn.
ESCAP (Economic and Social Commission for Asia and the Pacific) 1992: A Preliminary Study on Environmental Management of Tourism Development in the ESCAP Region. United Nations, New York.
EUROPARAT 1975: Richtlinie des Rates vom 8.12. 1975 über die Qualität von Badegewässern [Council of Europe, Directive of the Council of Dec. 8, 1975 on the Quality of Bathing Waters] (76/160/EWG).
EUROPARAT 1993: Verordnung (EWG) über die freiwillige Beteiligung gewerblicher Unternehmen an einem Gemeinschaftssystem für das Umweltmanagement und die Umweltbetriebsprüfung [Directive (EEC) on Voluntary Participation of Business Enterprises in a Community System for Environmental Management and the Environmental Company Audit], Brussels.
EUROPEAN ENVIRONMENT AGENCY (EEA) 1995: Europe's Environment, The Dobris Assessment. Ed. by D. Stanners and P. Bourdeau. European Environment Agency, Copenhagen (unpublished draft).
EUROPÄISCHE UMWELTAGENTUR [European Environment Agency] (n.d.): CORINE Biotopes Manual – Habitats of the European Community, Brussels.
EUROPEAN UNION FOR COASTAL CONSERVATION (EUCC) 1991: Sand Dune Inventory of Europe, Leiden, Netherlands.

EUCC 1992: Die Küste: Eine große Aufgabe für Europa [The Coast: A Great Task for Europe], Leiden, Netherlands.

EUCC 1994: Coastal Ecosystems and Tourism – A Study of the Environmental Impacts of Different Forms of Tourism (draft), Leiden, Netherlands.

FAIRBRIDGE, R.W. (ed.) 1966: Encyclopedia of Oceanography. Encyclopedia of Earth Science Series, vol. I, New York.

FAIRBRIDGE, R.W. (ed.) 1968: Encyclopedia of Geomorphology. Encyclopedia of Earth Science Series, vol. III, New York.

FEDERATION OF NATURE AND NATIONAL PARKS OF EUROPE (FNNPE) 1993: Loving Them to Death? Sustainable Tourism in Europe's Nature and National Parks. Kliemo, Eupen, Belgium.

FISCHER, M. 1993: [various brief articles and translations], in: Tranvia, no. 30.

FONTAUBERT, C. de 1993: Earth Summit Aftermath – Creation of the Commission on Sustainable Development. In: Ocean & Coastal Management, vol. 20, 1993 (special Issue: Ecotourism in Marine and Coastal Areas).

FRANK, D. 1992: Einwirkung touristischer Aktivitäten in der Nachsaison auf die Rastvögel im Deichvorland eines Fremdenverkehrsortes. Gutachten erstellt im Auftrag der Nationalparkverwaltung Niedersächsisches Wattenmeer [Impact of Touristic Activities in the Postseason on Resting Birds in the Foreshore of a Tourist Town. Expert Opinion on Commission of the National Park Administration of the Lower Saxon Wadden Sea], Wilhelmshaven.

FREMDENVERKEHRSWIRTSCHAFT (FVW), Dec. 27, 1995: Dr. Peter Aderhold: Moderates Wachstum auch in den nächsten Jahren [Moderate Growth in the Coming Years, as Well].

FRITZ, G. 1992: Leitgedanken und Typen der Tourismusentwicklung unter besonderer Berücksichtigung der Umwelt- und Naturverträglichkeit [Central Themes and Types of Touristic Development, with Special Consideration of Environmental Sustainability], Bonn.

FUCHS, O./KOLBE, C./MARTENS, D./SCHLEIFNECKER, T. 1993: Konzept für einen angepaßten und nachhaltigen Tourismus im Refugio Nacional de Vida Silvestre Gandoca-Manzanillo, Costa Rica. Projektarbeit am Institut für Landesplanung und Raumforschung [Programme for an Adapted and Sustainable Tourism in the Vida Silvestre Gandoca-Manzanillo National Refuge, Costa Rica. Project Report at the Institute for Land Planning and Regional Research], University of Hannover, Hannover.

GEOJOURNAL LIBRARY, THE 1993: Tourism vs Environment: The Case for Coastal Areas. Edited by P. P. Wong. Kluwer Academic Publishers, Dordrecht/ Boston/London.

GIERLOFF-EMDEN, H.G. 1980: Geographie des Meeres, Ozeane und Küsten [Geography of the Seas, Oceans and Coasts], vol. 2 (Lehrbuch der Allgemeinen Geographie), Berlin/New York.

GLASSNER, M.I. 1993: Management of Marine Resources as a Binding Force in the Eastern Caribbean. In: Ocean & Coastal Management 20 (1993), Essex, U.K.

GLOWKA, L./BURHENNE-GUILMIN, F./SYNGE, H./McNEELY, J./GÜNDLING, L. 1994: A Guide to the Convention on Biological Diversity. Environmental Policy and Law. Paper no. 30 (publ. by IUCN Environmental Law Centre, IUCN Biodiversity Programme), Gland and Cambridge.

GOMEZ, M.M. et al. 1992: Ocio y Turismo con los parques naturales Andaluses [Leisure Time and Tourism in the Nature Parks of Andalusia]. Serie Documentos; Turismo no. 1, publ. by Dirección General de Turismo. Consejera de Economica y hacienda. Junta de Andalucia, p. 73.

GORMSEN, E. 1987a: Der Fremdenverkehr in Lateinamerika und seine Folgen für Regionalstruktur und kulturellen Wandel [Tourism in Latin America and Its Consequences for Regional Structure and Cultural Change]. In: Lateinamerika im Brennpunkt, Berlin.

GORMSEN, E. 1987b: Der Tourismus und seine Folgen für Mensch und Umwelt in Lateinamerika [Tourism and Its Consequences for Man and the Environment in Latin America]. In: Tübinger Geographische Studien no. 96, p. 241–252, Tübingen.

GREENPEACE ESPAÑA 1993: Abwassereinleitungen ins Mittelmeer [Sewage Discharge into the Mediterranean]. In: Tranvia, no. 30, p. 26f.

GRENON, M./BATISSE, M. 1991: Futures for the Mediterranean Basin: The Blue Plan. Oxford University Press, Oxford, U.K.

GROVES, D.G./HUNT, L.M. 1980: The Ocean World Encyclopedia, New York.

GÜNDLING, L. 1995: Biologische Vielfalt - genetische Ressourcen - Technologietransfer. Über Ökologie und Ökonomie in der Konvention über die biologische Vielfalt [Biological Diversity, Genetic Resources, Technology Transfer: On Biology and Economy in the Convention on Biological Diversity], in: Loccumer Protokolle 66/94, 1995, p. 63ff.

GUNN, C.A. 1994: Tourism Planning: Basis, Concepts, Cases, 3rd ed., Taylor & Francis Washington, Bristol, London.

HAAS, P.M. 1990: Saving the Mediterranean. The Politics of International Environmental Cooperation. Columbia University Press, New York.

HÄLKER, M. A. 1995: Andalusien. DuMont Buchverlag, Cologne.

HALL, M./WOUTERS, M. 1994: Managing Nature Tourism in the Sub-Antarctic. In: Annals of Tourism Research, vol. 21, no. 2.

HAMELE, H. 1993: Datenbank für eine dauerhafte und umweltgerechte Tourismusentwicklung [Data Bank for a Lasting and Sustainable Tourism Development]. ECOTRANS. In: Politische Ökologie, no. 32, July/August, 1993, Munich.

HARFST, W./SCHARPF H. 1983: Fremdenverkehrsbedingte Umweltbelastungen, Pilotstudie [Environmental Stress Caused by Tourism: Pilot Study], Hannover.

HAWARD, M. 1993: Ocean and Coastal Zone Management in Australia and New Zealand: Current Initiatives. In: Ocean & Coastal Management 19 (1993), Recent Developments and Announcements, Essex, U.K.

HELCOM 1994: 20 Years of International Cooperation for the Baltic Marine Environment 1974–1994. Helsinki.

HELFER, M. 1993: Tourismus auf Rügen. Chancen und Risiken der Umstrukturierung infolge der deutschen Einigung [Tourism on Rügen: Opportunities and Risks of Restructuring Following German Unification]. Arbeiten aus dem geographischen Institut der Universität des Saarlandes, vol. 40, Saarbrücken.

HERRMANN, S./LAMBRECHT, H./WAGNER, R. 1990: Feriengroßprojekte unter UVP-Gesichtspunkten - Das Fallbeispiel "Center Parc Bispingen" [Large-Scale Holiday Projects Under EIA Aspects: The Example of "Bispingen Center Parc"]. Reihe Arbeitsmaterialien des Institutes für Landschaftspflege und Naturschutz am Fachbereich Landespflege der Universität Hannover, vol. 13, Hannover.

HOERING, U. 1993: Das Trojanische Pferd vor dem Weltbank-Tor - Viel mehr als Umweltkosmetik kam beim Rio-Fonds bisher nicht heraus [The Trojan Horse ante Portas of the World Bank: Much More than Environmental Cosmetics Has Thus Far Not Happened with the Rio Fund]. In: Frankfurter Rundschau, Dec. 7, 1993.

HOLST, J. 1994: Bauboom an der Küste - In Chile verschandeln immer mehr Betonburgen für Touristen die wunderschöne Landschaft [Building Boom on the Coast: In Chile More and More Concrete Palaces Are an Eyesore for Tourists in the Beautiful Landscape]. In: die tageszeitung, Nov. 19, 1994, Berlin.

HOLTHUS, P. 1993: Recent Developments in the South Pacific. In: Ocean & Coastal Management 19 (1993), Recent Developments and Announcements, Essex, U.K.

HORIZONTES NATURE TOURS 1993: Costa Rican Tourism Tomorrow - Where Do We Go from Here? The Barceló Controversy. In: Horizontes Nature Tours News, vol. 2, no. 1, San José, Costa Rica.

ICARD, M./ZIPPRICK, J. 1994: Bordeaux und Atlantikküste [Bordeaux and the Atlantic Coast]. DuMont Buchverlag, Cologne.

ICT (Instituto Costarricense de Turismo) 1993: Costa Rica, tourism press bulletin, vol. 1 (1993), no. 1 to 3, April–December, 1993.

ICT (1994): [Data on touristic development in Costa Rica].

IDEAL 1995: El litoral andaluz afronta el verano con el 96% de las playas en buen estado hygienico. La junta dice que las playas sin bandera azul tambien poseen gran calidad y exelentes condiciones [The Andalusian Coast Is Approaching the Summer with 96% of the Beaches in Good Hygienic Condition. The Commission Says that the Beches Without Blue Flags Are Also of Fine Quality and Excellent Condition], in: Granada (regional newspaper of Andalusia), June, 10 and 19, 1995.

INSKEEP, E. 1992: Sustainable Tourism Development in the Maldives and Bhutan. In: UNEP 1992.

INSKEEP, E./KALLENBERGER, M. 1992: An Integrated Approach to Resort Development. Six Case Studies. Publ. by WTO, Madrid.

ISEP (International Society for Environmental Protection) 1992: Strategies for Reducing Environmental Impact of Tourism. Proceedings of the ENVIROTOUR Vienna, November 10–12, 1992.

ISHMAEL, L. 1991: Ecotourism: The Planning Issues. Paper presented at the First Caribbean Conference on Ecotourism, July 9–12, in Belize City, Belize.

ISHMAEL, L. 1993: Environmental Impacts from Cruise Tourism. In: CTO 1993.

INSTITUT FRANÇAIS DE L'ENVIRONNEMENT (ifen) 1994: L'Environnement en France. Edition 1994–1995. Paris.

IUCN (International Union for Conservation of Nature) 1991: Legal Aspects of the Conservation of Wetlands, Gland, Cambridge, U.K.

IUCN 1993a: Nature Reserves of the Himalaya and Mountains of Central Asia. Prepared by the World Conservation Monitoring Centre. IUCN, Gland, Switzerland and Cambridge, U.K.

IUCN 1994a: Guidelines for Protected Area Management Categories. CNPPA with the assistance of WCMC. IUCN, Gland, Switzerland and Cambridge, U.K.

IUCN 1994b: 1993 United Nations List of National Parks and Protected Areas. Prepared by WCMC and CNPPA. IUCN, Gland, Switzerland and Cambridge, U.K.

IUCN/BID 1993b: Parques y progreso – Areas Protegidas y Desarrollo Económico en América Latina y el Caribe [Parks and Progress: Protected Areas and Economic Development in Latin America and the Caribbean]. IUCN Cambridge, U.K.

IUCN/UNEP/WWF 1991: Unsere Verantwortung für die Erde – Strategie für ein Leben in Einklang mit Natur und Umwelt [Our Responsibility for the Earth: Strategy for a Life in Harmony with Nature and the Environment], Gland.

JACKSON, I. 1986: Carrying Capacity for Tourism in Small Tropical Caribbean Islands. In: Industry and Environment, vol. 9 (1986) no. 1, Nairobi.

JACKSON, M.H. 1989: Conservation in the Islands. In: Galápagos – a Natural History Guide. The University of Calgary Press.

JOB, H. 1994: Der Nationalpark als regionaler Entwicklungsfaktor? – Eine vorläufige Analyse am Beispiel "Kisiti-Mpunguti Marine National Park" und Wasini Island (Kenia) [The National Park as a Factor of Regional Development? – An Interim Analysis of the Example of the Kaisiti-Mpunguti Marine National Park and Wasini Island, Kenya]. Manuscript, Trier.

JOURNAL OF SUSTAINABLE TOURISM 1993: Reports: Tourism and the Environment: Challenges and Choices for the 1990s, vol. 1, no. 1.

KASPAR, R. 1992: Ski Tourism in New Zealand: Toward Sustainable Management. In: ISEP 1992.

KELLETAT, D. 1993: Coastal Geomorphology and Tourism on the German North Sea Coast. In: Wong 1993.

KENCHINGTON, R. A. 1990: Tourism in Coastal and Marine Environments: A Recreational Perspective. In: MILLER/AUYONG 1991.

KERN, M. 1995: Ansätze zu einer nachhaltigen Tourismuspolitik des Bundes und der Länder [Approaches to a Sustainable Tourism Policy of the Federal Government and the States], vol. 31, Trier.

KING, B./WEAVER, S. 1993: The Impact of the Environment on the Fiji Tourism Industry: A Study of Industry Attitudes. In: Journal of Sustainable Tourism, vol. 1, no. 2.

KLEINKE, J. 1992: National- und Naturparkführer Mecklenburg-Vorpommern [Guide to National and Nature Parks in Mecklenburg-Vorpommern]. Demmler Verlag, Schwerin.

KNAPP, H.D. et al. 1986: Die Landschaftsgeschichte der Insel Rügen seit dem Spätglazial [The History of the Landscape of the Isle of Rügen since the Late Glacial Period]. Schriften zur Ur- und Frühgeschichte. Akademie Verlag Berlin.

KNAPP, H.D. et al. 1988: Landschaftsgeschichte als interdisziplinare Arbeitsrichtung, dargestellt am Beispiel der Insel Rügen [Landscape History as an Interdisciplinary Field as Exemplified by the Isle of Rügen], Jena.

KNAPP, H.D. 1995: Was ist biologische Vielfalt? [What Is Biological Diversity?] In: Biologische Vielfalt erhalten! Eine Aufgabe der Entwicklungszusammenarbeit, Dokumentation zum Fachgespräch "Umsetzung der Biodiversitätskonvention" am 13.6.1995 in Eschborn [Preserve Biological Diversity! A Task of Development Cooperation: Documentation on the Expert Discussion on "Implementation of the Biodiversity Convention" on June 13, 1995 in Eschborn].

KNECHT, R.W./CICIN-SAIN, B. 1993: Earth Summit Held: Stage Set for New Global Partnership. In: Ocean & Coastal Management 19 (1993), Recent Developments and Announcements, Essex, U.K.

KNOKE, V./STOCK, M. 1994: Menschliche Aktivitäten im Schleswig-Holsteinischen Wattenmeer und deren Auswirkungen auf Vögel [Human Activities in the Schleswig-Holstein Wadden Sea and Their Impact on Birds]. Wattenmeerstelle Husum der Umweltstiftung WWF-Deutschland, Husum.

KOESTER, V. 1989: The Ramsar Convention on the Conservation of Wetlands, Gland

KÖRBER, S. 1995: Schlechte Noten für Flüsse und Seen. EU-Kommission prüfte Badequalität der Badegewässer [Bad Marks for Rivers and Lakes: The EU Commission Tested Bathing Quality of Bathing Waters]. In: Berliner Morgenpost, June 17, 1995, Berlin.

KRAFT, F. 1994: Wasser-Test. Küsten, die noch sauber sind [Water Test: Coasts Which Are Still Clean]. In: Stern 26/1994, Hamburg.

KULINAT, K. 1986: Fremdenverkehr in Spanien [Tourism in Spain]. In: Geographische Rundschau, vol. 38, no.1, p. 28–35.

KUTAY, K. 1991: Cahuita National Park, Costa Rica: A Case Study in Living Cultures and National Park Management. In: WEST/BRECHIN 1991, Part Four: Ecodevelopment Based on Sustained Yield Local Resource Utilization.

L'ENVIRONNEMENT EN FRANCE: Rapport sur l'etat de l'environnement en France [Report on the State of the Environment in France], Dunod et Institut Français de l'environnement, Paris.

LAFONT, J./ZYSBERG, C. 1992: Tourisme et environnement – l'expérience française [Tourism and the Environment: French Experiences]. In: UNEP 1992.

LAMP, J. 1988: Ökologische Probleme durch Tourismus an der Nordseeküste [Ecological Problems Through Tourism on the North Sea Coast]. In: Wattenmeer International 4/88.

LAMP, J./FRICKE, H. 1989: Sanfter Tourismus – eine Chance für die Küste [Soft Tourism: an Opportunity for the Coast]; Tagungsbericht 3 der Umweltstiftung WWF-Deutschland. Naturfreunde Verlag Freizeit und Wandern GmbH, Stuttgart.

LA ROCA , F. 1993: Krieg um Wasser? [War over Water?] In: Travania, no. 30, p. 18f.

LEA, J. 1988: Tourism and Development in the Third World. Routledge, London/New York.

LESER, H. et al. 1984: Diercke Wörterbuch der Allgemeinen Geographie [Diercke Dictionary of General Geography], 2 vols., Braunschweig.

LIEDL et al. 1992: Die Ostsee. Meeresnatur im ökologischen Notstand [The Baltic Sea: Marine Nature in an Ecological Emergency], Göttingen.

LINDBERG, K. (1994): Quantifying Ecotourism – Are Reliable Statistics in Sight? In: The Ecotourism Society Newsletter, vol. 4, no. 2, spring, 1994.

LIPMAN, G.H. 1992: Travel and Tourism – Bridge Between Environment and Development. In: UNEP 1992.

LORCH, L. et al. 1995: Nachhaltige Entwicklung im Alpenraum [Sustainable Development in the Alpine Region], Umweltbundesamt, Texte 15/1995, Berlin.

MacGREGOR, J. 1993: Planegap – Planification de l'écotourisme et gestion des aires potentiels. An Integrated Planning Model for Ecotourism Product Development in Madagascar. (Publ. by Association Nationale pour la Gestion des Aires Protégés Madagascar and The Ecotourism Society).

MARAJH, O./MEADOWS, D.R. 1992: Ecotourism in Latin America and the Caribbean – Strategies and Implications for Development. Manuscript of a paper presented at Envirotour Vienna, 1992, East Lansing, Michigan, USA.

MARINELLIS, C. 1993: Die Power kommt vom Dach. Echt Öko auf St. John: Das Resort Harmony zeigt, wie man es macht [The Power Comes from the Roof. Pure Eco on St. John: The Harmony Resort Shows How to Do It]. In: touristik aktuell, no. 44, Nov. 2, 1993, p. 28.

MILLER, M. 1993: The Rise of Coastal and Marine Tourism. In: Ocean & Coastal Management, vol. 20, 1993 (special issue: Ecotourism in Marine and Coastal Areas).

MILLER, M./AUYONG, J. (eds.) 1991a: Proceedings of the 1990 Congress on Coastal and Marine Tourism. A Symposium and Workshop on Balancing Conservation and Economic Development, Honolulu, Hawaii, USA, May 25–31, 1990.

MILLER, M./AUYONG, J. 1991b: Coastal Zone Tourism – A Potent Force Affecting Environment and Society. In: Marine Policy, March, 1991.

MILLER, M./DITTON, R. 1986: Travel, Tourism, and Marine Affairs. In: Coastal Zone Management Journal, vol. 14, no. 1/2.

MIOSSEC, A. 1993: Tourist Development and Coastal Conservation in France. In: Wong 1993, p. 167–187.

MÜLLER, H. 1993: Im Dschungel der Zeichen. Ökologische Produktdeklaration im Tourismus [In the Jungle of Symbols: Ecological Product Declaration in Tourism]. In: Politische Ökologie, no. 32, July/August, 1993, Munich.

MÜLLER, H. 1994: The Thorny Path to Sustainable Tourism Development. In: Journal of Sustainable Tourism, vol. 2, no. 3, 1994.

NIEUWENHUIS, J. 1992: Policies with Respect to Tourism and the Environment. In: WTO 1992b.

NORDBERG, L. 1994: Examples of Legislation on Coastlines: Outlook for Pan-European Harmonisation? (Finland), in: Colloquy on "The Protection of Coastal Areas of the Adriatic Sea". Council of Europe, Strasbourg.

ÖKOSYSTEMFORSCHUNG WATTENMEER 1992: Berichte [Wadden Sea Ecosystem Research, Reports], no. 1/1992. 2. Wissenschaftliches Symposium, March 4–5, 1991, Büsum.

ÖKOSYSTEMFORSCHUNG WATTENMEER 1994: Eine Zwischenbilanz [An Interim Report]. Schriftenreihe, no. 5. Tönning.

ÖKOSYSTEMFORSCHUNG (Schleswig-Holsteinisches) WATTENMEER 1994a: Menschliche Aktivitäten im Schleswig-Holsteinischen Wattenmeer und deren Auswirkungen auf Vögel [Human Activities in the Wadden Sea of Schleswig-Holstein and Their Impacts on Birds]. Husum.

OLDENBURG, K. 1988: Umweltbelastung und Landschaftsplanung in der Türkei [Environmental Stress and Landscape Planning in Turkey]. In: Garten und Landschaft 12/88, Munich.

PAHR, W. 1987: Im Interesse der Gäste wie der Gastgeber – Die Weltorganisation für Tourismus (WTO) [In the Interest of the Guests and Their Hosts: The World Tourism Organization]. In: Vereinte Nationen 3/87.

PEARCE, D. G./KIRK, R. M. 1986: Carrying Capacities for Coastal Tourism. In: UNEP (ed.): Industry and Environment, vol. 9, no. 1.

PERNETTA, J.C./ELDER, D.L. 1993: Cross-sectoral, Integrated Coastal Area Planning (CICAP): Guidelines and Principles for Coastal Area Development. IUCN, Gland, Switzerland.

PRICE, A.R.G./HEINANEN, A.P./GIBSON, J.P./YOUNG, E.R. 1992: Guidelines for Developing a Coastal Zone Management Plan for Belize. IUCN, Gland, Switzerland.

PRICE, A.R.G./HUMPHREY, S.L. 1993: Application of the Biosphere Reserve Concept to Coastal Marine Areas: Papers presented at the UNESCO/IUCN San Francisco Workshop of 14–20 August 1989. IUCN, Gland, Switzerland.

PROGNOS AG 1994: Leitbilder und Ziele für eine umweltschonende Raumentwicklung in der Ostsee-Küstenregion Mecklenburg-Vorpommerns [Guidelines and Goals for Sustainable Regional Development in the Baltic Coastal Region of Mecklenburg-Vorpommern], publ. by Umweltbundesamt, Berlin.

RÄTH, B. 1993: Tourismus-Gütesiegel Grüner Koffer. Die Mühlsteine der Interessen [The Green Suitcase Tourism Quality Seal: The Millstones of Interests]. In: Politische Ökologie, no. 32, July/August, 1993, Munich.

RAMSAMY, S. 1992: Tourism Development and the Environment at Island Destinations – The Example of Mauritius. In: UNEP 1992.

RATHNAM, M./OPSAL, K. 1989: Preparation of an Environmental Action Plan for Mauritius. In: Industry and Environment (publ. by UNEP), July–December, 1989.

RATHS, U./RIECKEN, U./SSYMANK, A. 1995: Gefährdung von Lebensraumtypen in Deutschland und ihre Ursachen – Auswertung der Roten Listen gefährdeter Biotoptypen [Threats to Habitat Types in Germany and Their Causes]. In: Natur und Landschaft, vol. 70. (1995), no. 5, Stuttgart.

RICHARDSON, J. 1993: Australia Takes Sustainable Tourism Route. In: The Ecotourism Society Newsletter, vol. 3, no. 2 (spring, 1993). Bennington, USA.

RILEY, C.W. 1990: Bermuda's Cruise Ship Industry – Headed for the Rocks? In: MILLER/AUYONG 1991a.

RIVERA, A. 1993: Das Feuchtgebiet Doñana in Andalusien [The Doñana Wetlands in Andalusia]. In: Taverna, no. 30, p. 16f.

SALM, R.V. 1986: Coral Reefs and Tourist Carrying Capacity – the Indian Ocean Experience. In: Industry and Environment, vol. 9 (1986), no. 1. Nairobi.

SALMAN, A. 1994: Colloquium on 'The Protection of Coastal Areas of the Adriatic Sea'. Prospects for the Conservation of Coastal Areas, Strasbourg.

SANDBERG, B. 1994: Côte d'Azur. DuMont Buchverlag, Köln.

SANSON, L. 1994: An Ecotourism Case Study in Sub-Antarctic Islands.

SATHIENDRAKUMAR, R./TISDELL, C.A. 1990: Marine Areas as Tourist Attractions in the Southern Indian Ocean. In: MILLER/AUYONG 1991.

SCHARPF, H. 1989: Steuerungsinstrumente zur umweltgerechten Entwicklung des Tourismus im Nationalpark-Umfeld an der Küste [Instruments for Controlling Sustainable Development of Tourism in the Environs of National Parks on the Coast]. In: Festschrift für Konrad Buchwald zum 75. Geburtstag, TU Hannover, p. 255–275.

SCHEERENS, J. 1993: Glacier Bay Under Pressure – Alaska's Cruise Vessel Dilemma. In: The Ecotourism Society Newsletter, vol. 3, no. 3 (summer, 1993), Bennington, USA.

SCHEMEL, H.J./ERBGUTH, W. 1992: Handbuch Sport und Umwelt [Sports and Environment Handbook]. Meyer & Meyer Verlag, Aachen.

SCHEMEL, H.J./RUHL, G. 1980: Umweltverträgliche Planung im Alpenraum – Die Zusammenhänge zwischen Nutzungsansprüchen und Umweltressourcen in den deutschen Alpen [Sustainable Planning in the Alpine Region: The Correlation of Use Demand and Environmental Resources in the German Alps]. Studie i. A. des Bundesministers des Innern und des Umweltbundesamtes Berlin, Berlin.

SCHERB, K. 1975: Die Abwasserbeseitigung auf Campingplätzen [Sewage Disposal at Camping Grounds]. In: Wasser für die Erholungslandschaft, Münchner Beiträge zur Abwasser-Fischerei und Flußbiologie, vol. 26, publ. by Bayerische Versuchsanstalt München. Munich.

SCHERFOSE, V./YÜCEL, M. 1988: Das geplante Hotelbauprojekt in Köyegiz/Dalyan an der türkischen Südküste und seine ökologischen Folgen [The Planned Hotel Construction Project in Köyegiz/Dayan on the Turkish Southern Coast and Its Ecological Consequences]. In: UVP-Report 3/88, Dortmund.

SCHMIDT, G. 1995: Naturschutzplanung in Spanien. Verbindung zwischen Naturschutz und umweltverträglicher Entwicklung – ein Modell für die ländlichen Räume Europas [Conservation Planning in Spain: A Nexus Between Conservation and Sustainable Development – A Model for the Rural Areas of Europe]. In: Naturschutz und Landschaftsplanung 27, (2), 1995, Stuttgart.

SCHMIDT, N./LUDWIG, A. 1993: Keine Zeit für Unendlichkeitsgedanken – Auf Fraser Island in Australiens Osten wird sanfter Tourismus straff organisiert [No Time to Think About Eternity: On Fraser Island in Australia's East Soft Tourism Is Strictly Organised]. In: Frankfurter Rundschau, Dec. 31, 1993.

SCHNEIDER, U. 1994: Touristisches Entwicklungskonzept für die Gemeinde Ummanz auf Rügen [Touristic Development Programme for the Community of Ummanz on Rügen (unpubl. thesis)]. Fachbereich 7, Umwelt und Gesellschaft, TU Berlin, Berlin.

SCHOLZ, W. 1993: Hochfliegende Pläne und düstere Aussichten – Tourismusplanungen auf Sansibar [High-flying Plans and Somber Prospects: Tourism Planning on Zanzibar]. In: HÄUSLER et al. 1993.

SCHÜMER, R. 1993: Einsatzmöglichkeiten von Geographischen Informationssystemen bei der UVP [Possibilities for Using Geographic-Information Systems in an EIA], thesis, TU Berlin, Berlin.

SHACKLEFORD, P. 1995: personal communication, Jan. 10, 1995.

SINGH, T.V. 1992: Development of Tourism in the Himalayan Environment – The Problem of Sustainability. In: UNEP (1992).

SINGH, T.V./THEUNS, H.L./GO, F.M. (eds.) 1988: Towards Appropriate Tourism: The Case of Developing Countries. Europäische Hochschulschriften, series X, Fremdenverkehr [Tourism], vol. 11, Frankfurt/M.

SMITH, R.I.L./WALTON, D.W.H./DINGWALL, P.R. 1994: Developing the Antarctic Protected Area System. IUCN, Gland, Switzerland and Cambridge, U.K.

SMITH, V.L. 1994: A Sustainable Antarctic – Science and Tourism. In: Annals of Tourism Research, vol. 21, no. 2, 1994.

SOLBRIG, O.T. 1994: Biodiversität – Wissenschaftliche Fragen und Vorschläge für die internationale Forschung [Biodiversity: Scientific Questions and Proposals for International Research]. Publ. by Deutsches Nationalkomitee für das UNESCO-Programm "Der Mensch und die Biosphäre" (MAB), Bonn.

SPEHS, P. 1990: Neue staatlich geplante Badeorte in Mexiko [New Government-planned Bathing Resorts in Mexico]. In: Geographische Rundschau 42 (1990), no. 1.

SPITTLER, R. 1996: Tourismus und Naturschutz auf Nord-Rügen. Möglichkeiten und Grenzen einer naturschutzverträglichen Tourismusentwicklung [Tourism and Conservation on Northern Rügen: Possibilities and Limitations of Conservation-Compatible Tourism Development]. In: Arbeitsberichte der Arbeitsgemeinschaft Geographie Münster e.V., vol. 26. Münster.

SPLETTSTOESSER, J./FOLKS, M.C. 1994: Environmental Guidelines for Tourism in Antarctica. In: Annals of Tourism Research, vol. 21, no. 2, 1994.

SSYMANK, A. 1994: Neue Anforderungen im europäischen Naturschutz. Das Schutzgebietssystem Natura 2000 und die "FFH-Richtlinie" der EU [New Challenges in European Conservation: The Nature 2000 Protected-Area System and the "FFH Directive" of the EU]. In Natur und Landschaft, vol. 69 (1994), no. 9, Stuttgart.

STACHOWITSCH, M. 1992: Tourism and the Sea – The World's Largest Ecosystem in Danger. In: ISEP 1992.

STATISTISCHES BUNDESAMT 1995: Tourismus in Zahlen 1994 [Tourism Statistics, 1994]. Statistisches Bundesamt, Wiesbaden.

STERR, H. 1993: Auseinandersetzung mit dem globalen Wandel in Küstenregionen [Meeting the Challenge of Global Change in Coastal Regions]. In: Global Change Prisma vol. 4 (4).

STEWART, M. C. 1993: Sustainable Tourism Development and Marine Conservation Regimes. In: Ocean & Coastal Management, vol. 20, 1993 (special issue: Ecotourism in Marine and Coastal Areas).
STOCK, M./SCHULZ, R. 1991: Seeregenpfeifer und Touristen in St. Peter-Böhl. Konflikte im Nationalpark Schleswig-Holsteinisches Wattenmeer [Plovers and Tourists in St. Peter-Böhl: Conflicts in the Schleswig-Holstein Wadden Sea National Park]. In: Wattenmeer International, 1/91.
STORM, P.-C./LOHSE, S. n.d.: EG-Umweltrecht, Loseblattsammlung [EC Environmental Law, loose-leaf edition], Berlin.
STRASDAS, W. 1994: Auswirkungen neuer Freizeittrends auf die Umwelt [Impacts of New Leisure-time Trends on the Environment], Meyer u. Meyer Verlag, Berlin.
SUMMERER 1995: Schritte zu einem nachhaltigen Deutschland [Steps to a Sustainable Germany], Lecture delivered at the TU Berlin, Nov. 1, 1995. Berlin.

TALAMANCA ASSOCIATION FOR ECOTOURISM AND CONSERVATION (ATEC) n. d.: Coastal Talamanca – a Cultural and Ecological Guide, Puerto Viejo, Talamanca.
TARNAS, D. A./HASSAN, K. 1990: Sustainable Development in Marine Tourism for South Johor, Malaysia – The ASEAN-USAID Coastal Resources Management Project. In: MILLER/AUYONG 1991.
TAVERNE, B. 1995: Planning and Management Tools for Establishment of Sustainable Tourism. Demonstrated in a Coastal Zone of the Wadden Sea. In: Tourismus und Umwelt in Europa, Brussels/Luxemburg 1995, p. 38–40.
TES (THE ECOTOURISM SOCIETY) 1993a: Ecotourism Guidelines for Nature Tour Operators, N. Bennington, USA.
TES (ed.) 1993b: Ecotourism: A Guide for Planners and Managers, N. Bennington, USA.
TES 1993c: Trash Dumping by Cruise Lines Is Threatening Caribbean Waters. In: The Ecotourism Society Newsletter, vol. 3, no. 1 (winter, 1993), Bennington, USA.
TES 1993d: Galapagos Tourism Plan Progresses. In: The Ecotourism Society Newsletter, vol. 3, no. 1 (winter, 1993), Bennington, USA.
TES 1993e: Costa Rica Expotur Controversy. In: The Ecotourism Society Newsletter, vol. 3, no. 3 (summer, 1993), Bennington, USA.
TES 1993f: International Update: U.S. Coast Guard Reports on Cruise Ships. In: The Ecotourism Society Newsletter, vol. 3, no. 3 (summer, 1993), Bennington, USA.
TES 1994: Coral Reef Protection Initiative Launched. In: The Ecotourism Society Newsletter, vol. 4, no. 3 (summer, 1994), Bennington, USA.
THOMAS, P. 1990: Coastal and Marine Tourism – A Conservation Perspective. In: MILLER/AUYONG 1991.
TKALEK, M. 1995: Ein neuer Beruf: Umwelt-Vermittler. Arbeitsgemeinschaft Umwelt diskutiert über Konfliktregelung auch Streitfall "Naturpark Rügen" [A New Profession: Environmental Arbitrator. The Environment Study Group Discusses Resolving Conflicts in the Dispute over "Rügen Nature Park"]. In: Berliner Zeitung, no. 248, Oct. 24, 1995.

TÖDTER, U. 1992: Die Alpenkonvention – Argumente der CIPRA zum Regelungsbedarf im Alpentourismus [The Alps Convention: Arguments of the CIPRA on the Need for Regulating Alpine Tourism]. In: ISEP 1992.

TROTTNO, B. 1992: Naturtourismus als Mogelpackung – Die ökologischen Probleme sind in Dalyan nicht weniger geworden [Nature Tourism as Phoney Packaging: The Ecological Problems Have Not Been Reduced in Dalyan]. In: Frankfurter Rundschau, May 16, 1992.

TRUMBIC, I. 1994: Coastal Area Management Programme in Albania. In: Colloquy on the Protection of Coastal Areas of the Adriatic Sea, Strasbourg.

TU BERLIN 1989: Projektbericht 1989. Sylt und Föhr. Zwei Inseln und ihre Probleme. Studienprojekt am Fachbereich 14, Landschaftsplanung. TU Berlin [Project Report for 1989: Sylt and Föhr, Two Islands and Their Problems. Study project in landscape planning at the TU Berlin].

TUI (Touristik Union International) 1993: Urlaub und Umwelt – Informationen für den TUI-Service zu Umweltbelastungen und Lösungsansätzen in den Zielgebieten [Holiday and Environment: Information for TUI Service on Environmental Stress and Solution Strategies in Destinations], Hannover.

TÜRKER, A. 1991: Das neue Reiseziel: Türkei. Soziale und ökologische Auswirkung des Massentourismus in einem Schwellenland [The New Travel Destination – Turkey: Social and Ecological Impacts of Mass Tourism in a Threshold Country], Berlin.

TÜRKER, A./FAHRENHORST, B./SCHEUMANN, W. 1988: Umweltschutz und Tourismusentwicklung an der Türkischen Südwestküste [Environmental Protection and Tourism Development on the Turkish Southwest Coast]. Werkstattberichte des Instituts für Landschaftsökonomie, no. 23. Technische Universität Berlin, Berlin.

TZOANOS, G. 1992: Tourism and the Environment – The Role of the European Community. In: UNEP 1992.

UNEP (IE/PAC) 1992: Industry and Environment: Sustainable Tourism Development.

UNEP/SOUTH PACIFIC REGIONAL ENVIRONMENT PROGRAMME/IUCN/ASIAN DEVELOPMENT BANK/AUSTRALIAN CENTRE FOR ENVIRONMENTAL LAW 1992: Strengthening Environmental Legislation in the South Pacific Region. Workshop Materials. 23–27 November 1992, Apia, Western Samoa.

UNEP 1995: Global Biodiversity Assessment. Cambridge University Press.

UNEP IE (United Nations Environment Programme Industry and Environment) 1995: Environmental Codes of Conduct for Tourism. Technical Report no. 29, Paris.

VALLEGA, A. 1994: Mediterranean Action Plan – Which Futures? In: Ocean & Coastal Management, vol. 23 1994, Ireland, p. 271–279.

VAN'T HOF, T. 1993: Developing Management Systems for Marine Parks. In: CTO 1993.

VISSER, N./NJUGUNA, S. 1992: Environmental Impacts of Tourism on the Kenya Coast. In: UNEP 1992.
VORLAUFER, K. 1990: Dritte-Welt-Tourismus – Vehikel der Entwicklung oder Weg in die Unterentwicklung? [Third-World Tourism: Vehicle of Development or a Road to Underdevelopment?] In: Geographische Rundschau, vol. 42, 1990.

WADE, R. 1993: Leading the Way in Environmentally Sensitive Cruise Tourism. In: CTO 1993.
WALLACE, G. 1993: Visitor Management: Lessons from Galápagos National Park. In: LINDBERG/HAWKINS 1993.
WASCHER, D. 1993: Die Bedeutung des CORINE-Biotop-Projektes und der geplanten Europäischen Umweltagentur für den Naturschutz in der Europäischen Gemeinschaft [The Importance of the CORINE Biotope Project and the Planned European Environmental Agency for Conservation in the European Community]. In: Natur und Landschaft, vol. 68, (1993), no. 3, Stuttgart.
WASCHER, D. 1995: Vielfalt durch Vernetzung [Diversity Through Networking]. In: Politische Ökologie 43, Nov./Dec., 1995, p. 53–56.
WCMC (WORLD CONSERVATION MONITORING CENTRE) 1992: Global Biodiversity – Status of the Earth's Living Resources, Chapman & Hall, London.
WCMC 1994: Biodiversity Data Sourcebook. Ed. by B. Groombridge and M. Jenkins. World Conservation Press, Cambridge, U.K.
WEISS, I. 1991: Am Korallenriff blinken die Coladosen – Tauchtouristen schwimmen auf den Malediven im eigenen Müll [On the Coral Reef Are Shiny Coke Cans: Skindiving Tourists on the Maldives Are Swimming in Their Own Garbage]. In: Frankfurter Rundschau, Apr. 27, 1991.
WICHMANN, W. 1994: 'Bitte nicht füttern'. Die Ostsee erstickt an Nährstoffen ['Please Do Not Feed': The Baltic Is Suffocating from Nutrients]. In: Berliner Morgenpost, Jan. 9, 1994.
WILKINSON, C.R./BUDDEMEIER, R.W. 1994: Global Climate Change and Coral Reefs: Implications for People and Reefs. IUCN, Gland, Switzerland.
WISSENSCHAFTLICHER BEIRAT DER BUNDESREGIERUNG 1995: Welt im Wandel. Wege zur Lösung globaler Umweltprobleme. Jahresgutachten [Scientific Advisory Panel of the Federal Government, World in Transition: Ways of Solving Global Environmental Problems, Annual Report], Springer Verlag, Berlin/Heidelberg.
WONG, P.P. 1993: Tourism vs Environment: The Case for Coastal Areas. Kluwer Academic Publishers, Dordrecht/Boston/London.
WTO (WORLD TOURISM ORGANIZATION) 1990: Seminar on the Integration of Tourism in Europe, Istanbul, Turkey, 7–9 May 1990. World Tourism Organization, Madrid, Spain.
WTO 1990: Compendium of Tourism Statistics 1985–1989. 11th ed., Madrid.
WTO 1991: Tourism to the Year 2000 – Qualitative Aspects Affecting Global Growth. A Discussion Paper (Executive Summary), Madrid.

WTO 1992: Seminar on the Tourism Sector and Construction of the European Community, Bruges (Belgium), 5–6 May 1992. World Tourism Organization, Madrid, Spain.

WTO 1993: Sustainable Tourism Development – Guide for Local Planners. A Tourism and the Environment Publication, Madrid.

WTO 1994a: Compendium of Tourism Statistics 1988–1992. 14th ed., Madrid.

WTO 1994b: National and Regional Tourism Planning – Methodologies and Case Studies. Routledge, London/New York.

WTO/UNEP (UNITED NATIONS ENVIRONMENT PROGRAMME) (eds.) 1992: Guidelines – Development of National Parks and Protected Areas for Tourism. WTO, UNEP-IE/PAC Technical Report Series, no. 13, Madrid/Paris.

WTO NEWS 1994: No. 1–5.

WTTC (World Travel and Tourism Council) n.d.: Green Globe – A Worldwide Environmental Program for the Travel & Tourism Industry, London.

WTTC n.d.: The World Travel & Tourism Council – Background (Media information), Brüssel.

WWF (World Wide Fund for Nature) 1992: Beyond the Green Horizon – A Discussion Paper on Principles for Sustainable Tourism, Godalming, Surrey, U.K.

WWF-Projektbüro Wattenmeer (ed.) 1995: Wattenmeer International, vol. 13, no. 1. Husum.

ZIMMER, D. 1995: Dürre, hausgemacht [Homemade Drought]. In: ZEIT magazin, no. 30, p. 6–12.

MIX
Papier aus verantwortungsvollen Quellen
Paper from responsible sources
FSC® C105338
FSC
www.fsc.org

If you have any concerns about our products,
you can contact us on
ProductSafety@springernature.com

In case Publisher is established outside the EU,
the EU authorized representative is:
Springer Nature Customer Service Center GmbH
Europaplatz 3, 69115 Heidelberg, Germany

Printed by Libri Plureos GmbH
in Hamburg, Germany